LLVM Techniques, Tips, and Best Practices Clang and Middle-End Libraries

Design powerful and reliable compilers using
the latest libraries and tools from LLVM

Min-Yih Hsu

BIRMINGHAM—MUMBAI

LLVM Techniques, Tips, and Best Practices Clang and Middle-End Libraries

Copyright © 2021 Packt Publishing

Group Product Manager: Aaron Lazar
Publishing Product Manager: Shweta Bairoliya
Senior Editor: Nitee Shetty
Content Development Editor: Kinnari Chohan
Technical Editor: Pradeep Sahu
Copy Editor: Safis Editing
Project Coordinator: Deeksha Thakkar
Proofreader: Safis Editing
Indexer: Tejal Daruwale Soni
Production Designer: Jyoti Chauhan

First published: April 2021

Production reference: 1220421

Published by Packt Publishing Ltd.
Livery Place
35 Livery Street
Birmingham
B3 2PB, UK.

ISBN 978-1-83882-495-2

www.packt.com

To my parents.

– Min-Yih Hsu

Contributors

About the author

Min-Yih "Min" Hsu is a Ph.D. candidate in computer science at the University of California, Irvine. His research focuses on compiler engineering, code optimization, advanced hardware architectures, and system security. He has been an active member of the LLVM community since 2015 and has contributed numerous patches upstream. Min also dedicates his time to advocating LLVM and compiler engineering through various avenues, such as writing blog posts and giving talks at conferences. In his spare time, Min likes to explore a variety of different coffee beans and ways of brewing.

I want to thank the people who have supported me, especially my family and my academic advisors. I also want to thank the LLVM community for always being inclusive and kind to every member regardless of their background and origin.

About the reviewer

Suyog Sarda completed his B.Tech from the College of Engineering, Pune. His work so far has mostly been related to compilers. He is especially interested in the performance aspect of compilers. He has worked on domain-specific language for image processing for DSPs.

LLVM's modularity makes it interesting to learn and implement quickly according to the compiler's requirement. However, the documentation of LLVM is scattered. He hopes this book provides a consolidated overview of the LLVM compiler infrastructure.

Table of Contents

3

Testing with LLVM LIT

4

TableGen Development

Section 2: Frontend Development

5

Exploring Clang's Architecture

6

Extending the Preprocessor

7

Handling AST

8

Working with Compiler Flags and Toolchains

Section 3: "Middle-End" Development

9
Working with PassManager and AnalysisManager

10
Processing LLVM IR

11

Gearing Up with Support Utilities

12

Learning LLVM IR Instrumentation

Assessments

Other Books You May Enjoy

Index

Preface

A compiler is one of the most prevailing tools used by programmers. The majority of programmers have compilers – or some form of compilation technique – in their development flow. A modern compiler not only transforms high-level programming languages into low-level machine code, but also plays a key role in optimizing the speed, size, or even the memory footprint of the program it compiles. With these characteristics, building a production-ready compiler has always been a challenging task.

LLVM is a framework for compiler optimization and code generation. It provides building blocks that significantly reduce the efforts of developers to create high-quality optimizing compilers and programming language tools. One of its most famous products is Clang – the C-family language compiler that builds thousands of pieces of widely-used software including the Google Chrome browser and iOS apps. LLVM is also used in compilers for many different programming languages, such as the famous Swift programming language. It is not an exaggeration to say that LLVM is one of the hottest topics when it comes to creating a new programming language.

With hundreds of libraries and thousands of different APIs, LLVM provides a wide range of features, from key functionalities for optimizing a program to more general utilities. In this book, we provide a complete and thorough developer guide to two of the most important sub-systems in LLVM – Clang and the middle-end. We start with introductions to several components and development best practices that can benefit your general development experiences with LLVM. Then, we will show you how to develop with Clang. More specifically, we will focus on the topics that help you augment and customize the functionalities in Clang. In the last part of this book, you will learn crucial knowledge about LLVM IR development. This includes how to write an LLVM Pass with the latest syntax and mastering processing different IR constructions. We also show you several utilities that can greatly improve your productivity in LLVM development. Last but not least, we don't assume any particular LLVM version in this book – we try to keep up to date and include the latest features from the LLVM source tree.

This book provides a handful of code snippets and sample projects in every chapter. You are encouraged to download them from the GitHub repository of this book and play around with your own customizations.

Who this book is for

This book is for people of all LLVM experience levels, with a basic understanding of compilers. If you are a compiler engineer who uses LLVM in your daily work, this book provides concise development guidelines and references. If you are an academic researcher, this book will help you learn useful LLVM skills and build your prototypes and projects in a short time. Programming language enthusiasts will also find this book useful when it comes to building a new programming language with the help of LLVM.

What this book covers

Chapter 1, Saving Resources When Building LLVM, gives a brief introduction to the LLVM project, before showing you how to build LLVM without draining your CPU, memory resources, and disk space. This paves the road to shorter development cycles and smoother experiences for later chapters.

Chapter 2, Exploring LLVM's Build System Features, shows you how to write CMake build scripts for both in-tree and out-of-tree LLVM development. You will learn crucial skills to leverage LLVM's custom build system features to write more expressive and robust build scripts.

Chapter 3, Testing with LLVM LIT, shows you the way to run testing with LLVM's LIT infrastructure. The chapter not only gives you a better understanding of how testing works in LLVM's source tree but also enables you to integrate this intuitive, scalable testing infrastructure into any project.

Chapter 4, TableGen Development, shows you how to write TableGen – a special **Domain Specific Language (DSL)** invented by LLVM. We especially focus on using TableGen as a general tool for processing structural data, giving you flexible skills to use TableGen outside LLVM.

Chapter 5, Exploring Clang's Architecture, marks the start of our topics on Clang. This chapter gives you an overview of Clang, especially its compilation flow, and presents to you the role of individual components in Clang's compilation flow.

Chapter 6, Extending the Preprocessor, shows you the architecture of the preprocessor in Clang and, more importantly, shows you how to develop a plugin to extend its functionalities without modifying any code in the LLVM source tree.

Chapter 7, Handling AST, shows you how to develop with an **Abstract Syntax Tree (AST)** in Clang. The content includes learning important topics to work with an AST's in-memory representation and a tutorial to create a plugin that inserts custom AST processing logic into the compilation flow.

Chapter 8, Working with Compiler Flags and Toolchains, covers the steps to add custom compiler flags and toolchains to Clang. Both skills are especially crucial if you want to support new features or new platforms in Clang.

Chapter 9, Working with PassManager and AnalysisManager, marks the start of our discussion on the LLVM middle-end. This chapter focuses on writing an LLVM pass – using the latest new PassManager syntax – and how to access program analysis data via AnalysisManager.

Chapter 10, Processing LLVM IR, is a big chapter containing a variety of core knowledge regarding LLVM IR, including the structure of LLVM IR's in-memory representation and useful skills to work with different IR units such as functions, instructions, and loops.

Chapter 11, Gearing Up with Support Utilities, introduces some utilities that can improve your productivity – such as having better debugging experiences – when working with LLVM IR.

Chapter 12, Learning LLVM IR Instrumentation, shows you how instrumentation works on LLVM IR. It covers two primary use cases: Sanitizer and **Profile-Guided Optimization (PGO)**. For the former, you will learn how to create a custom sanitizer. For the latter, you will learn how to leverage PGO data in your LLVM Pass.

To get the most out of this book

This book is designed to bring you the latest features of LLVM, so we encourage you to use LLVM after version 12.0, or even the development branch – that is, the main branch – throughout this book.

We assume that you are working on Linux or Unix systems (including macOS). Tools and sample commands in this book are mostly run in the command-line interface, but you are free to use any code editors or IDEs to write your code.

Software/hardware covered in the book	OS requirements
LLVM >= version 12.x	macOS or Linux (Any distribution)
GCC >= version 5.1, or any C/C++ compiler that supports C++14 standard	
CMake >= version 3.13.4	
Python 3.x	
Ninja Build or GNU Make	
Graphviz	

In *Chapter 1, Saving Resources on Building LLVM*, we will provide details on how to build LLVM from source.

If you are using the digital version of this book, we advise you to type the code yourself or access the code via the GitHub repository (link available in the next section). Doing so will help you avoid any potential errors related to the copying and pasting of code.

Download the example code files

You can download the example code files for this book from GitHub at `https://github.com/PacktPublishing/LLVM-Techniques-Tips-and-Best-Practices-Clang-and-Middle-End-Libraries`. In case there's an update to the code, it will be updated on the existing GitHub repository.

We also have other code bundles from our rich catalog of books and videos available at `https://github.com/PacktPublishing/`. Check them out!

Download the color images

We also provide a PDF file that has color images of the screenshots/diagrams used in this book. You can download it here: `https://static.packt-cdn.com/downloads/9781838824952_ColorImages.pdf`.

Conventions used

There are a number of text conventions used throughout this book.

`Code in text`: Indicates code words in text, database table names, folder names, filenames, file extensions, pathnames, dummy URLs, user input, and Twitter handles. Here is an example: "To include Clang in the build list, please edit the value assigned to the `LLVM_ENABLE_PROJECTS` CMake variable."

A block of code is set as follows:

```
TranslationUnitDecl 0x560f3929f5a8 <<invalid sloc>> <invalid
sloc>
|...
`-FunctionDecl 0x560f392e1350 <./test.c:2:1, col:30> col:5 foo
'int (int)'
```

When we wish to draw your attention to a particular part of a code block, the relevant lines or items are set in bold:

```
   |-ParmVarDecl 0x560f392e1280 <col:9, col:13> col:13 used c
'int'
   `-CompoundStmt 0x560f392e14c8 <col:16, col:30>
     `-ReturnStmt 0x560f392e14b8 <col:17, col:28>
       `-BinaryOperator 0x560f392e1498 <col:24, col:28> 'int'
'+'
          |-ImplicitCastExpr 0x560f392e1480 <col:24> 'int'
<LValueToRValue>
          | `-DeclRefExpr 0x560f392e1440 <col:24> 'int' lvalue
ParmVar 0x560f392e1280 'c' 'int'
          `-IntegerLiteral 0x560f392e1460 <col:28> 'int' 1
```

Any command-line input or output is written as follows:

```
$ clang -fplugin=/path/to/MyPlugin.so … foo.cpp
```

Bold: Indicates a new term, an important word, or words that you see onscreen. For example, words in menus or dialog boxes appear in the text like this. Here is an example: "Select **System info** from the **Administration** panel."

> **Tips or important notes**
> Appear like this.

Get in touch

Feedback from our readers is always welcome.

General feedback: If you have questions about any aspect of this book, mention the book title in the subject of your message and email us at customercare@packtpub.com.

Errata: Although we have taken every care to ensure the accuracy of our content, mistakes do happen. If you have found a mistake in this book, we would be grateful if you would report this to us. Please visit www.packtpub.com/support/errata, selecting your book, clicking on the Errata Submission Form link, and entering the details.

Piracy: If you come across any illegal copies of our works in any form on the Internet, we would be grateful if you would provide us with the location address or website name. Please contact us at `copyright@packt.com` with a link to the material.

If you are interested in becoming an author: If there is a topic that you have expertise in and you are interested in either writing or contributing to a book, please visit `authors.packtpub.com`.

Reviews

Please leave a review. Once you have read and used this book, why not leave a review on the site that you purchased it from? Potential readers can then see and use your unbiased opinion to make purchase decisions, we at Packt can understand what you think about our products, and our authors can see your feedback on their book. Thank you!

For more information about Packt, please visit `packt.com`.

Section 1: Build System and LLVM-Specific Tooling

You will learn the advanced skills of developing LLVM's build system for both in-tree and out-of-tree scenarios. This section includes the following chapters:

- *Chapter 1, Saving Resources When Building LLVM*
- *Chapter 2, Exploring LLVM's Build System Features*
- *Chapter 3, Testing with LLVM LIT*
- *Chapter 4, TableGen Development*

1
Saving Resources When Building LLVM

LLVM is the state-of-the-art compiler optimization and code generation framework adopted by many amazing industrial and academic projects, such as the **Just-In-Time (JIT)** compiler in JavaScript engines and **machine learning (ML)** frameworks. It is a useful toolbox for building programming languages and binary file tools. However, despite the project's robustness, its learning resources are scattered, and it doesn't have the best documentation either. Due to this, it has a pretty steep learning curve, even for developers with some LLVM experience. This book aims to tackle these issues by providing you with knowledge of common and important domains in LLVM in a pragmatic fashion – showing you some useful engineering tips, pointing out lesser-known but handy features, and illustrating useful examples.

As an **LLVM** developer, building LLVM from source has always been the first thing you should do. Given the scale of LLVM nowadays, this task can take hours to finish. Even worse, rebuilding the project to reflect changes might also take a long time and hinder your productivity. Therefore, it's crucial to know how to use the right tools and how to find the best build configurations for your project for the sake of saving various resources, especially your precious time.

In this chapter, we are going to cover the following topics:

- Cutting down building resources with better tooling
- Saving building resources by tweaking CMake arguments
- Learning how to use GN, an alternative LLVM build system, and its pros and cons

Technical requirements

At the time of writing this book, LLVM only has a few software requirements:

- A C/C++ compiler that supports C++14
- CMake
- One of the build systems supported by CMake, such as GNU Make or Ninja
- Python (2.7 is fine too, but I strongly recommend using 3.x)
- zlib

The exact versions of these items change from time to time. Check out `https://llvm.org/docs/GettingStarted.html#software` for more details.

This chapter assumes you have built an LLVM before. If that's not the case, perform the following steps:

1. Grab a copy of the LLVM source tree from GitHub:

   ```
   $ git clone https://github.com/llvm/llvm-project
   ```

2. Usually, the default branch should build without errors. If you want to use release versions that are more stable, such as release version 10.x, use the following command:

   ```
   $ git clone -b release/10.x https://github.com/llvm/llvm-project
   ```

3. Finally, you should create a build folder where you're going to invoke the CMake command. All the building artifacts will also be placed inside this folder. This can be done using the following command:

   ```
   $ mkdir .my_build
   $ cd .my_build
   ```

Cutting down building resources with better tooling

As we mentioned at the beginning of this chapter, if you build LLVM with the default (CMake) configurations, by invoking **CMake** and building the project in the following way, there is a high chance that the whole process will take *hours* to finish:

```
$ cmake ../llvm
$ make all
```

This can be avoided by simply using better tools and changing some environments. In this section, we will cover some guidelines to help you choose the right tools and configurations that can both speed up your building time and improve memory footprints.

Replacing GNU Make with Ninja

The first improvement we can do is using the **Ninja** build tool (`https://ninja-build.org`) rather than GNU Make, which is the default build system generated by CMake on major Linux/Unix platforms.

Here are the steps you can use to set up Ninja on your system:

1. On Ubuntu, for example, you can install Ninja by using this command:

    ```
    $ sudo apt install ninja-build
    ```

    Ninja is also available in most Linux distributions.

2. Then, when you're invoking CMake for your LLVM build, add an extra argument:

    ```
    $ cmake -G "Ninja" ../llvm
    ```

3. Finally, use the following build command instead:

    ```
    $ ninja all
    ```

Ninja runs *significantly* faster than GNU Make on large code bases such as LLVM. One of the secrets behind Ninja's blazing fast running speed is that while the majority of build scripts such as `Makefile` are designed to be written manually, the syntax of Ninja's build script, `build.ninja`, is more similar to assembly code, which should *not* be edited by developers but generated by other higher-level build systems such as CMake. The fact that Ninja uses an assembly-like build script allows it to do many optimizations under the hood and get rid of many redundancies, such as slower parsing speeds, when invoking the build. Ninja also has a good reputation for generating better dependencies among build targets.

Ninja makes clever decisions in terms of its *degree of parallelization*; that is, how many jobs you want to execute in parallel. So, usually, you don't need to worry about this. If you want to explicitly assign the number of worker threads, the same command-line option used by GNU Make still works here:

```
$ ninja -j8 all
```

Let's now see how you can avoid using the BFD linker.

Avoiding the use of the BFD linker

The second improvement we can do is using linkers *other than* the BFD linker, which is the default linker used in most Linux systems. The BFD linker, despite being the most mature linker on Unix/Linux systems, is not optimized for speed or memory consumption. This would create a performance bottleneck, especially for large projects such as LLVM. This is because, unlike the compiling phase, it's pretty hard for the linking phase to do file-level parallelization. Not to mention the fact that the BFD linker's peak memory consumption when building LLVM usually takes about 20 GB, causing a burden on computers with small amounts of memory. Fortunately, there are at least two linkers in the wild that provide both good single-thread performance and low memory consumption: the **GNU gold linker** and LLVM's own linker, **LLD**.

The gold linker was originally developed by Google and donated to GNU's `binutils`. You should have it sitting in the `binutils` package by default in modern Linux distributions. LLD is one of LLVM's subprojects with even faster linking speed and an experimental parallel linking technique. Some of the Linux distributions (newer Ubuntu versions, for example) already have LLD in their package repository. You can also download the prebuilt version from LLVM's official website.

To use the gold linker or LLD to build your LLVM source tree, add an extra CMake argument with the name of the linker you want to use.

For the gold linker, use the following command:

```
$ cmake -G "Ninja" -DLLVM_USE_LINKER=gold ../llvm
```

Similarly, for LLD, use the following command:

```
$ cmake -G "Ninja" -DLLVM_USE_LINKER=lld ../llvm
```

> **Limiting the number of parallel threads for Linking**
>
> Limiting the number of parallel threads for linking is another way to reduce (peak) memory consumption. You can achieve this by assigning the `LLVM_PARALLEL_LINK_JOBS=<N>` CMake variable, where N is the desired number of working threads.

With that, we've learned that by simply using different tools, the building time could be reduced *significantly*. In the next section, we're going to improve this building speed by tweaking LLVM's CMake arguments.

Tweaking CMake arguments

This section will show you some of the most common CMake arguments in LLVM's build system that can help you customize your build and achieve maximum efficiency.

Before we start, you should have a build folder that has been CMake-configured. Most of the following subsections will modify a file in the build folder; that is, `CMakeCache.txt`.

Choosing the right build type

LLVM uses several predefined build types provided by CMake. The most common types among them are as follows:

- `Release`: This is the default build type if you didn't specify any. It will adopt the highest optimization level (usually -O3) and eliminate most of the debug information. Usually, this build type will make the building speed slightly slower.

- `Debug`: This build type will compile without any optimization applied (that is, -O0). It preserves all the debug information. Note that this will generate a *huge* number of artifacts and usually take up ~20 GB of space, so please be sure you have enough storage space when using this build type. This will usually make the building speed slightly faster since no optimization is being performed.

- `RelWithDebInfo`: This build type applies as much compiler optimization as possible (usually -O2) and preserves all the debug information. This is an option balanced between space consumption, runtime speed, and debuggability.

You can choose one of them using the `CMAKE_BUILD_TYPE` CMake variable. For example, to use the `RelWithDebInfo` type, you can use the following command:

```
$ cmake -DCMAKE_BUILD_TYPE=RelWithDebInfo …
```

It is recommended to use `RelWithDebInfo` first (if you're going to debug LLVM later). Modern compilers have gone a long way to improve the debug information's quality in optimized program binaries. So, always give it a try first to avoid unnecessary storage waste; you can always go back to the `Debug` type if things don't work out.

In addition to configuring build types, `LLVM_ENABLE_ASSERTIONS` is another CMake (Boolean) argument that controls whether assertions (that is, the `assert(bool predicate)` function, which will terminate the program if the predicate argument is not true) are enabled. By default, this flag will only be true if the build type is `Debug`, but you can always turn it on manually to enforce stricter checks, even in other build types.

Avoiding building all targets

The number of LLVM's supported targets (hardware) has grown rapidly in the past few years. At the time of writing this book, there are nearly 20 officially supported targets. Each of them deals with non-trivial tasks such as native code generation, so it takes a significant amount of time to build. However, the chances that you're going to be working on *all* of these targets at the same time are low. Thus, you can select a subset of targets to build using the `LLVM_TARGETS_TO_BUILD` CMake argument. For example, to build the X86 target only, we can use the following command:

```
$ cmake -DLLVM_TARGETS_TO_BUILD="X86" …
```

You can also specify multiple targets using a semicolon-separated list, as follows:

```
$ cmake -DLLVM_TARGETS_TO_BUILD="X86;AArch64;AMDGPU" …
```

> **Surround the list of targets with double quotes!**
>
> In some shells, such as BASH, a semicolon is an ending symbol for a command. So, the rest of the CMake command will be cut off if you don't surround the list of targets with *double-quotes*.

Let's see how building shared libraries can help tweak CMake arguments.

Building as shared libraries

One of the most iconic features of LLVM is its **modular design**. Each component, optimization algorithm, code generation, and utility libraries, to name a few, are put into their own libraries where developers can link individual ones, depending on their usage. By default, each component is built as a **static library** (`*.a` in Unix/Linux and `*.lib` in Windows). However, in this case, static libraries have the following drawbacks:

- Linking against static libraries usually takes more time than linking against dynamic libraries (`*.so` in Unix/Linux and `*.dll` in Windows).

- If multiple executables link against the same set of libraries, like many of the LLVM tools do, the total size of these executables will be *significantly* larger when you adopt the static library approach compared to its dynamic library counterpart. This is because each of the executables has a copy of those libraries.

- When you're debugging LLVM programs with debuggers (GDB, for example), they usually spend quite some time loading the statically linked executables at the very beginning, hindering the debugging experience.

Thus, it's recommended to build every LLVM component as a dynamic library during the development phase by using the `BUILD_SHARED_LIBS` CMake argument:

```
$ cmake -DBUILD_SHARED_LIBS=ON …
```

This will save you a significant amount of storage space and speed up the building process.

Splitting the debug info

When you're building a program in debug mode – adding the `-g` flag when using you're GCC and Clang, for example – by default, the generated binary contains a section that stores **debug information**. This information is essential for using a debugger (for example, GDB) to debug that program. LLVM is a large and complex project, so when you're building it in debug mode – using the `cMAKE_BUILD_TYPE=Debug` variable – the compiled libraries and executables come with a huge amount of debug information that takes up a lot of disk space. This causes the following problems:

- Due to the design of C/C++, several *duplicates* of the same debug information might be embedded in different object files (for example, the debug information for a header file might be embedded in every library that includes it), which wastes lots of disk space.

- The linker needs to load object files AND their associated debug information into memory during the linking stage, meaning that memory pressure will increase if the object file contains a non-trivial amount of debug information.

To solve these problems, the build system in LLVM provides allows us to *split* debug information into separate files from the original object files. By detaching debug information from object files, the debug info of the same source file is condensed into one place, thus avoiding unnecessary duplicates being created and saving lots of disk space. In addition, since debug info is not part of the object files anymore, the linker no longer needs to load them into memory and thus saves lots of memory resources. Last but not least, this feature can also improve our *incremental* building speed – that is, rebuild the project after a (small) code change – since we only need to update the modified debug information in a single place.

To use this feature, please use the `LLVM_USE_SPLIT_DWARF` cmake variable:

```
$ cmake -DcmAKE_BUILD_TYPE=Debug -DLLVM_USE_SPLIT_DWARF=ON …
```

Note that this CMake variable only works for compilers that use the DWARF debug format, including GCC and Clang.

Building an optimized version of llvm-tblgen

TableGen is a **Domain-Specific Language** (**DSL**) for describing structural data that will be converted into the corresponding C/C++ code as part of LLVM's building process (we will learn more about this in the chapters to come). The conversion tool is called `llvm-tblgen`. In other words, the running time of `llvm-tblgen` will affect the building time of LLVM itself. Therefore, if you're not developing the TableGen part, it's always a good idea to build an optimized version of `llvm-tblgen`, regardless of the global build type (that is, `CMAKE_BUILD_TYPE`), making `llvm-tblgen` run faster and shortening the overall building time.

The following CMake command, for example, will create build configurations that build a debug version of everything *except* the `llvm-tblgen` executable, which will be built as an optimized version:

```
$ cmake -DLLVM_OPTIMIZED_TABLEGEN=ON -DCMAKE_BUILD_TYPE=Debug …
```

Lastly, you'll see how you can use Clang and the new PassManager.

Using the new PassManager and Clang

Clang is LLVM's official C-family frontend (including C, C++, and Objective-C). It uses LLVM's libraries to generate machine code, which is organized by one of the most important subsystems in LLVM – **PassManager**. PassManager puts together all the tasks (that is, the Passes) required for optimization and code generation.

In *Chapter 9*, *Working with PassManager and AnalysisManager*, will introduce LLVM's *new* PassManager, which builds from the ground up to replace the existing one somewhere in the future. The new PassManager has a faster runtime speed compared to the legacy PassManager. This advantage indirectly brings better runtime performance for Clang. Therefore, the idea here is pretty simple: if we build LLVM's source tree using Clang, with the new PassManager enabled, the compilation speed will be faster. Most of the mainstream Linux distribution package repositories already contain Clang. It's recommended to use Clang 6.0 or later if you want a more stable PassManager implementation. Use the LLVM_USE_NEWPM CMake variable to build LLVM with the new PassManager, as follows:

```
$ env CC=`which clang` CXX=`which clang++` \
    cmake -DLLVM_USE_NEWPM=ON …
```

LLVM is a huge project that takes a lot of time to build. The previous two sections introduced some useful tricks and tips for improving its building speed. In the next section, we're going to introduce an *alternative* build system to build LLVM. It has some advantages over the default CMake build system, which means it will be more suitable in some scenarios.

Using GN for a faster turnaround time

CMake is portable and flexible, and it has been battle-tested by many industrial projects. However, it has some serious issues when it comes to reconfigurations. As we saw in the previous sections, you can modify some of the CMake arguments once build files have been generated by editing the CMakeCache.txt file in the build folder. When you invoke the build command again, CMake will reconfigure the build files. If you edit the CMakeLists.txt files in your source folders, the same reconfiguration will also kick in. There are primarily two drawbacks of CMake's reconfiguration process:

- In some systems, the CMake configuration process is pretty slow. Even for reconfiguration, which theoretically only runs part of the process, it still takes a long time sometimes.

- Sometimes, CMake will fail to resolve the dependencies among different variables and build targets, so your changes will not reflect this. In the worst case, it will just silently fail and take you a long time to dig out the problem.

Generate Ninja, better known as **GN**, is a build file generator used by many of Google's projects, such as Chromium. GN generates Ninja files from its own description language. It has a good reputation for having a fast configuration time and reliable argument management. LLVM has brought GN support as an alternative (and experimental) building method since late 2018 (around version 8.0.0). GN is especially useful if your developments make changes to build files, or if you want to try out different building options in a short period.

Perform the following steps to use GN to build LLVM:

1. LLVM's GN support is sitting in the `llvm/utils/gn` folder. After switching to that folder, run the following `get.py` script to download GN's executable locally:

    ```
    $ cd llvm/utils/gn
    $ ./get.py
    ```

    > **Using a specific version of GN**
    >
    > If you want to use a custom GN executable instead of the one fetched by `get.py`, simply put your version into the system's `PATH`. If you are wondering what other GN versions are available, you might want to check out the instructions for installing `depot_tools` at `https://dev.chromium.org/developers/how-tos/install-depot-tools`.

2. Use `gn.py` in the *same* folder to generate build files (the local version of `gn.py` is just a wrapper around the real `gn`, to set up the essential environment):

    ```
    $ ./gn.py gen out/x64.release
    ```

    `out/x64.release` is the name of the build folder. Usually, GN users will name the build folder in `<architecture>.<build type>.<other features>` format.

3. Finally, you can switch into the build folder and launch Ninja:

    ```
    $ cd out/x64.release
    $ ninja <build target>
    ```

4. Alternatively, you can use the `-C` Ninja option:

    ```
    $ ninja -C out/x64.release <build target>
    ```

You probably already know that the initial build file generation process is super fast. Now, if you want to change some of the build arguments, please navigate to the `args.gn` file under the build folder (`out/x64.release/args.gn`, in this case); for example, if you want to change the build type to `debug` and change the targets to build (that is, the `LLVM_TARGETS_TO_BUILD` CMake argument) into `X86` and `AArch64`. It is recommended to use the following command to launch an editor to edit `args.gn`:

```
$ ./gn.py args out/x64.release
```

In the editor of `args.gn`, input the following contents:

```
# Inside args.gn
is_debug = true
llvm_targets_to_build = ["X86", "AArch64"]
```

Once you've saved and exited the editor, GN will do some syntax checking and regenerate the build files (of course, you can edit `args.gn` without using the gn command and the build files won't be regenerated until you invoke the `ninja` command). This regeneration/reconfiguration will also be fast. Most importantly, there won't be any infidelity behavior. Thanks to GN's language design, relationships between different build arguments can be easily analyzed with little ambiguity.

The list of GN's build arguments can be found by running this command:

```
$ ./gn.py args --list out/x64.release
```

Unfortunately, at the time of writing this book, there are still plenty of CMake arguments that haven't been ported to GN. GN is *not* a replacement for LLVM's existing CMake build system, but it is an *alternative*. Nevertheless, GN is still a decent building method if you want a fast turnaround time in your developments that involve many build configuration changes.

Summary

LLVM is a useful framework when it comes to building tools for code optimization and code generation. However, the size and complexity of its code base induces a non-trivial amount of build time. This chapter provided some tips for speeding up the build time of the LLVM source tree, including using different building tools, choosing the right CMake arguments, and even adopting a build system other than CMake. These skills cut down on unnecessary resource wasting and improve your productivity when developing with LLVM.

In the next chapter, we will dig into LLVM's CMake-based building infrastructure and show you how to build system features and guidelines that are crucial in many different development environments.

Further reading

- You can check out the complete list of CMake variables that are used by LLVM at `https://llvm.org/docs/CMake.html#frequently-used-CMake-variables`.

  You can learn more about GN at `https://gn.googlesource.com/gn`. The quick start guides at `https://gn.googlesource.com/gn/+/master/docs/quick_start.md are also very helpful`.

2
Exploring LLVM's Build System Features

In the previous chapter, we saw that LLVM's build system is a behemoth: it contains hundreds of build files with thousands of interleaving build dependencies. Not to mention, it contains targets that require custom build instructions for heterogeneous source files. These complexities drove LLVM to adopt some advanced build system features and, more importantly, a more structural build system design. In this chapter, our goal will be to learn about some important directives for the sake of writing more concise and expressive build files when doing both in-tree and out-of-tree LLVM developments.

In this chapter, we will cover the following main topics:

- Exploring a glossary of LLVM's important CMake directives
- Integrating LLVM via CMake in out-of-tree projects

Technical requirements

Similar to *Chapter 1*, *Saving Resources When Building LLVM*, you might want to have a copy of LLVM built from its source. Optionally, since this chapter will touch on quite a lot of CMake build files, you may wish to prepare a syntax highlighting plugin for `CMakeLists.txt` (for example, VSCode's *CMake Tools* plugin). All major IDEs and editors should have it off-the-shelf. Also, familiarity with basic `CMakeLists.txt` syntax is preferable.

All the code examples in this chapter can be found in this book's GitHub repository: `https://github.com/PacktPublishing/LLVM-Techniques-Tips-and-Best-Practices/tree/main/Chapter02`.

Exploring a glossary of LLVM's important CMake directives

LLVM has switched to **CMake** from **GNU autoconf** due to higher flexibility in terms of choosing underlying build systems. Ever since, LLVM has come up with many custom CMake functions, macros, and rules to optimize its own usage. This section will give you an overview of the most important and frequently used ones among them. We will learn how and when to use them.

Using the CMake function to add new libraries

Libraries are the building blocks of the LLVM framework. However, when writing `CMakeLists.txt` for a new library, you shouldn't use the normal `add_library` directive that appears in normal `CMakeLists.txt` files, as follows:

```
# In an in-tree CMakeLists.txt file…
add_library(MyLLVMPass SHARED
  MyPass.cpp) # Do NOT do this to add a new LLVM library
```

There are several drawbacks of using the vanilla `add_library` here, as follows:

- As shown in *Chapter 1*, *Saving Resources When Building LLVM,* LLVM prefers to use a global CMake argument (that is, `BUILD_SHARED_LIBS`) to control whether all its component libraries should be built statically or dynamically. It's pretty hard to do that using the built-in directives.

- Similar to the previous point, LLVM prefers to use a global CMake arguments to control some compile flags, such as whether or not to enable **Runtime Type Information** (**RTTI**) and **C++ exception handling** in the code base.
- By using custom CMake functions/macros, LLVM can create its own component system, which provides a higher level of abstraction for developers to designate build target dependencies in an easier way.

Therefore, you should always use the `add_llvm_component_library` CMake function shown here:

```
# In a CMakeLists.txt
add_llvm_component_library(LLVMFancyOpt
    FancyOpt.cpp)
```

Here, `LLVMFancyOpt` is the final library name and `FancyOpt.cpp` is the source file.

In regular CMake scripts, you can use `target_link_libraries` to designate a given target's library dependencies, and then use `add_dependencies` to assign dependencies among different build targets to create explicit build orderings. There is an easier way to do those tasks when you're using LLVM's custom CMake functions to create library targets.

By using the `LINK_COMPONENTS` argument in `add_llvm_component_library` (or `add_llvm_library`, which is the underlying implementation of the former one), you can designate the target's linked components:

```
add_llvm_component_library(LLVMFancyOpt
    FancyOpt.cpp
    LINK_COMPONENTS
    Analysis ScalarOpts)
```

Alternatively, you can do the same thing with the `LLVM_LINK_COMPONENTS` variable, which is defined before the function call:

```
set(LLVM_LINK_COMPONENTS
    Analysis ScalarOpts)
add_llvm_component_library(LLVMFancyOpt
    FancyOpt.cpp)
```

Component libraries are nothing but normal libraries with a special meaning when it comes to the *LLVM building blocks you can use*. They're also included in the gigantic libLLVM library if you choose to build it. The component names are slightly different from the real library names. If you need the mapping from component names to library names, you can use the following CMake function:

```
llvm_map_components_to_libnames(output_lib_names
   <list of component names>)
```

If you want to directly link against a *normal* library (the non-LLVM component one), you can use the LINK_LIBS argument:

```
add_llvm_component_library(LLVMFancyOpt
  FancyOpt.cpp
  LINK_LIBS
  ${BOOST_LIBRARY})
```

To assign general build target dependencies to a library target (equivalent to add_ dependencies), you can use the DEPENDS argument:

```
add_llvm_component_library(LLVMFancyOpt
  FancyOpt.cpp
  DEPENDS
  intrinsics_gen)
```

intrinsics_gen is a common target representing the procedure of generating header files containing LLVM intrinsics definitions.

Adding one build target per folder

Many LLVM custom CMake functions have a pitfall that involves source file detection. Let's say you have a directory structure like this:

```
/FancyOpt
   |___ FancyOpt.cpp
   |___ AggressiveFancyOpt.cpp
   |___ CMakeLists.txt
```

Here, you have two source files, FancyOpt.cpp and AggressiveFancyOpt.cpp. As their names suggest, FancyOpt.cpp is the basic version of this optimization, while AggressiveFancyOpt.cpp is an alternative, more aggressive version of the same functionality. Naturally, you will want to split them into separate libraries so that users can choose if they wish to include the more aggressive one in their normal workload. So, you might write a CMakeLists.txt file like this:

```
# In /FancyOpt/CMakeLists.txt
add_llvm_component_library(LLVMFancyOpt
  FancyOpt.cpp)
add_llvm_component_library(LLVMAggressiveFancyOpt
  AggressiveFancyOpt.cpp)
```

Unfortunately, this would generate error messages telling you something to the effect of Found unknown source AggressiveFancyOpt.cpp … when processing the first add_llvm_component_library statement.

LLVM's build system enforces a stricter rule to make sure that *all* C/C++ source files in the same folder are added to the same library, executable, or plugin. To fix this, it is necessary to split either file into a separate folder, like so:

```
/FancyOpt
  |___  FancyOpt.cpp
  |___  CMakeLists.txt
  |___  /AggressiveFancyOpt
        |___  AggressiveFancyOpt.cpp
        |___  CMakeLists.txt
```

In /FancyOpt/CMakeLists.txt, we have the following:

```
add_llvm_component_library(LLVMFancyOpt
  FancyOpt.cpp)
add_subdirectory(AggressiveFancyOpt)
```

Finally, in /FancyOpt/AggressiveFancyOpt/CMakeLists.txt, we have the following:

```
add_llvm_component_library(LLVMAggressiveFancyOpt
  AggressiveFancyOpt.cpp)
```

These are the essentials of adding build targets for (component) libraries using LLVM's custom CMake directives. In the next two sections, we will show you how to add executable and Pass plugin build targets using a different set of LLVM-specific CMake directives.

Using the CMake function to add executables and tools

Similar to `add_llvm_component_library`, to add a new executable target, we can use `add_llvm_executable` or `add_llvm_tool`:

```
add_llvm_tool(myLittleTool
  MyLittleTool.cpp)
```

These two functions have the same syntax. However, only targets created by `add_llvm_tool` will be included in the installations. There is also a global CMake variable, `LLVM_BUILD_TOOLS`, that enables/disables those LLVM tool targets.

Both functions can also use the `DEPENDS` argument to assign dependencies, similar to `add_llvm_library`, which we introduced earlier. However, you can only use the `LLVM_LINK_COMPONENTS` variable to designate components to link.

Using the CMake function to add Pass plugins

While we will cover Pass plugin development later in this book, adding a build target for a Pass plugin couldn't be any easier than now (compared to earlier LLVM versions, which were still using `add_llvm_library` with some special arguments). We can simply use the following command:

```
add_llvm_pass_plugin(MyPass
  HelloWorldPass.cpp)
```

The `LINK_COMPONENTS`, `LINK_LIBS`, and `DEPENDS` arguments are also available here, with the same usages and functionalities as in `add_llvm_component_library`.

These are some of the most common and important LLVM-specific CMake directives. Using these directives can not only make your CMake code more concise but also help synchronize it with LLVM's own build system, in case you want to do some in-tree development. In the next section, we will show you how to integrate LLVM into an out-of-tree CMake project, and leverage the knowledge we learned in this chapter.

> **In-tree versus out-of-tree development**
>
> In this book, *in-tree* development means contributing code directly to the LLVM project, such as fixing LLVM bugs or adding new features to the existing LLVM libraries. *Out-of-tree* development, on the other hand, either represents creating extensions for LLVM (writing an LLVM pass, for example) or using LLVM libraries in some other projects (using LLVM's code generation libraries to implement your own programming language, for example).

Understanding CMake integration for out-of-tree projects

Implementing your features in an in-tree project is good for prototyping, since most of the infrastructure is already there. However, there are many scenarios where pulling the entire LLVM source tree into your code base is not the best idea, compared to creating an **out-of-tree project** and linking it against the LLVM libraries. For example, you only want to create a small code refactoring tool using LLVM's features and open source it on GitHub, so telling developers on GitHub to download a multi-gigabyte LLVM source tree along with your little tool might not be a pleasant experience.

There are at least two ways to configure out-of-tree projects to link against LLVM:

- Using the `llvm-config` tool
- Using LLVM's CMake modules

Both approaches help you sort out all the details, including header files and library paths. However, the latter creates more concise and readable CMake scripts, which is preferable for projects that are already using CMake. This section will show the essential steps of using LLVM's CMake modules to integrate it into an out-of-tree CMake project.

First, we need to prepare an out-of-tree (C/C++) CMake project. The core CMake functions/macros we discussed in the previous section will help us work our way through this. Let's look at our steps:

1. We are assuming that you already have the following `CMakeLists.txt` skeleton for a project that needs to be linked against LLVM libraries:

```
project(MagicCLITool)
set(SOURCE_FILES
    main.cpp)
add_executable(magic-cli
  ${SOURCE_FILES})
```

Regardless of whether you're trying to create a project generating executable, just like the one we saw in the preceding code block, or other artifacts such as libraries or even LLVM Pass plugins, the biggest question now is how to get `include path`, as well as `library path`.

2. To resolve `include path` and `library path`, LLVM provides the standard CMake package interface for you to use the `find_package` CMake directive to import various configurations, as follows:

```
project(MagicCLITool)
find_package(LLVM REQUIRED CONFIG)
include_directories(${LLVM_INCLUDE_DIRS})
link_directories(${LLVM_LIBRARY_DIRS})
...
```

To make the `find_package` trick work, you need to supply the `LLVM_DIR` CMake variable while invoking the CMake command for this project:

```
$ cmake -DLLVM_DIR=<LLVM install path>/lib/cmake/llvm ...
```

Make sure it's pointing to the `lib/cmake/llvm` subdirectory under LLVM `install path`.

3. After resolving the include path and library, it's time to link the main executable against LLVM's libraries. LLVM's custom CMake functions (for example, `add_llvm_executable`) will be really useful here. But first, CMake needs to be able to *find* those functions.

    The following snippet imports LLVM's CMake module (more specifically, the `AddLLVM` CMake module), which contains those LLVM-specific functions/macros that we introduced in the previous section:

```
find_package(LLVM REQUIRED CONFIG)
...
list(APPEND CMAKE_MODULE_PATH ${LLVM_CMAKE_DIR})
include(AddLLVM)
```

4. The following snippet adds the executable build target using the CMake function we learned about in the previous section:

```
find_package(LLVM REQUIRED CONFIG)
...
include(AddLLVM)
set(LLVM_LINK_COMPONENTS
```

```
  Support
  Analysis)
add_llvm_executable(magic-cli
  main.cpp)
```

5. Adding the library target makes no difference:

```
find_package(LLVM REQUIRED CONFIG)
…
include(AddLLVM)
add_llvm_library(MyMagicLibrary
  lib.cpp
  LINK_COMPONENTS
  Support Analysis)
```

6. Finally, add the LLVM Pass plugin:

```
find_package(LLVM REQUIRED CONFIG)
…
include(AddLLVM)
add_llvm_pass_plugin(MyMagicPass
  ThePass.cpp)
```

7. In practice, you also need to be careful of **LLVM-specific definitions** and the RTTI setting:

```
find_package(LLVM REQUIRED CONFIG)
…
add_definitions(${LLVM_DEFINITIONS})
if(NOT ${LLVM_ENABLE_RTTI})
  # For non-MSVC compilers
  set(CMAKE_CXX_FLAGS "${CMAKE_CXX_FLAGS} -fno-rtti")
endif()
add_llvm_xxx(source.cpp)
```

This is especially true for the RTTI part because, by default, LLVM is not built with RTTI support, but normal C++ applications are. A compilation error will be thrown if there is an RTTI mismatch between your code and LLVM's libraries.

Despite the convenience of developing inside LLVM's source tree, sometimes, enclosing the entire LLVM source in your project might not be feasible. So, instead, we must create an out-of-tree project and integrate LLVM as a library. This section showed you how to integrate LLVM into your CMake-based out-of-tree projects and make good use of the LLVM-specific CMake directives we learned about in the *Exploring a glossary of LLVM's important CMake directives* section.

Summary

This chapter dug deeper into LLVM's CMake build system. We saw how to use LLVM's own CMake directives to write concise and effective build scripts, for both in-tree and out-of-tree development. Learning these CMake skills can make your LLVM development more efficient and provide you with more options to engage LLVM features with other existing code bases or custom logic.

In the next chapter, we will introduce another important infrastructure in the LLVM project known as the LLVM LIT, which is an easy-to-use yet general framework for running various kinds of tests.

3
Testing with LLVM LIT

In the previous chapter, we learned how to take advantage of LLVM's own CMake utilities to improve our development experience. We also learned how to seamlessly integrate LLVM into other out-of-tree projects. In this chapter, we're going to talk about how to get hands-on with LLVM's own testing infrastructure, LIT.

LIT is a testing infrastructure that was originally developed for running LLVM's regression tests. Now, it's not only the harness for running all the tests in LLVM (both **unit** and **regression tests**) but also a generic testing framework that can be used outside of LLVM. It also provides a wide range of testing formats to tackle different scenarios. This chapter will give you a thorough tour of the components in this framework and help you master LIT.

We are going to cover the following topics in this chapter:

- Using LIT in out-of-tree projects
- Learning about advanced FileCheck tricks
- Exploring the TestSuite framework

Technical requirements

The core of LIT is written in *Python*, so please make sure you have Python 2.7 or Python 3.x installed (Python 3.x is preferable, as LLVM is gradually retiring Python 2.7 now).

In addition, there are a bunch of supporting utilities, such as `FileCheck`, which will be used later. To build those utilities, the fastest way, unfortunately, is to build any of the `check-XXX` (phony) targets. For example, we could build `check-llvm-support`, as shown in the following code:

```
$ ninja check-llvm-support
```

Finally, the last section requires that `llvm-test-suite` has been built, which is a separate repository from `llvm-project`. We can clone it by using the following command:

```
$ git clone https://github.com/llvm/llvm-test-suite
```

The easiest way to configure the build will be using one of the cached CMake configs. For example, to build the test suite with optimizations (O3), we will use the following code:

```
$ mkdir .O3_build
$ cd .O3_build
$ cmake -G Ninja -DCMAKE_C_COMPILER=<desired Clang binary \
path> -C ../cmake/caches/O3.cmake ../
```

Then, we can build it normally using the following command:

```
$ ninja all
```

Using LIT in out-of-tree projects

Writing an in-tree LLVM IR regression test is pretty easy: all you need to do is annotate the IR file with testing directives. Look at the following script, for example:

```
; RUN: opt < %s -instcombine -S -o - | FileCheck %s
target triple = "x86_64-unknown-linux"
define i32 @foo(i32 %c) {
entry:
  ; CHECK: [[RET:%.+]] = add nsw i32 %c, 3
  ; CHECK: ret i32 [[RET]]
  %add1 = add nsw i32 %c, 1
```

```
  %add2 = add nsw i32 %add1, 2
  ret i32 %add2
}
```

This script checks if `InstCombine` (triggered by the `-instcombine` command-line option shown in the preceding snippet) simplifies two succeeding arithmetic adds into one. After putting this file into an arbitrary folder under `llvm/test`, the script will automatically be picked and run as part of the regression test when you're executing the `llvm-lit` command-line tool.

Despite its convenience, this barely helps you use LIT in out-of-tree projects. Using LIT out-of-tree is especially useful when your project needs some end-to-end testing facilities, such as a format converter, a text processor, a linter, and, of course, a compiler. This section will show you how to bring LIT to your out-of-tree projects, and then provide you with a complete picture of the running flow of LIT.

Preparing for our example project

In this section, we will use an out-of-tree CMake project. This example project builds a command-line tool, `js-minifier`, that *minifies* arbitrary JavaScript code. We will transform the following JavaScript code:

```
const foo = (a, b) => {
  let c = a + b;
  console.log(`This is ${c}`);
}
```

This will be transformed into some other *semantic-equivalent* code that is as short as possible:

```
const foo = (a,b) => {let c = a + b; console.log(`This is
${c}`);}
```

Instead of teaching you how to write this `js-minifier`, the goal of this section is to show you how to create a LIT testing environment to *test* this tool.

The example project has the following folder structure:

```
/JSMinifier
    |__ CMakeLists.txt
    |__ /src
        |__ js-minifier.cpp
```

```
|__ /test
   |__ test.js
   |__ CMakeLists.txt
|__ /build
```

The files under the /src folder contain the source code for js-minifier (which we are not going to cover here). What we will focus on here are the files that will be used for testing js-minifier, which sit under the /test folder (for now, there is only one file, test.js).

In this section, we are going to set up a testing environment so that when we run llvm-lit – the testing driver and main character of this section – under the CMake /build folder, it will print testing results, like this:

```
$ cd build
$ llvm-lit -sv .
-- Testing: 1 tests, 1 workers -
PASS: JSMinifier Test :: test.js (1 of 1)
Testing Time: 0.03s
  Expected Passes    : 1
```

This shows how many and what test cases have passed.

Here is the testing script, test.js:

```
// RUN: %jsm %s -o - | FileCheck

// CHECK: const foo = (a,b) =>
// CHECK-SAME: {let c = a + b; console.log(`This is ${c}`);}
const foo = (a, b) => {
  let c = a + b;
  console.log(`This is ${c}`);
}
```

As you can see, it is a simple testing process that runs the js-minifier tool – represented by the %jsm directive, which will be replaced by the real path to js-minifier executable, as explained later – and checks the running result with FileCheck by using its CHECK and CHECK-SAME directives.

With that, we've set up our example project. Before we wrap up the preparation, there is one final tool we need to create.

Since we're trying to cut down on our reliance on the LLVM source tree, recreate the `llvm-lit` command-line tool using the `LIT` package available in the *PyPi repository* (that is, the `pip` command-line tool). All you need to do is install that package:

```
$ pip install --user lit
```

Finally, wrap the package with the following script:

```
#!/usr/bin/env python
from lit.main import main
if __name__ == '__main__':
    main()
```

Now, we can use LIT without building an LLVM tree! Next, we will create some LIT configuration scripts that will drive the whole testing flow.

Writing LIT configurations

In this subsection, we'll show you how to write LIT configuration scripts. These scripts describe the testing process – where the files will be tested, the testing environment (if we need to import any tool, for example), the policy when there is a failure, and so on. Learning these skills can greatly improve how you use LIT in places outside the LLVM tree. Let's get started:

1. Inside the `/JSMinifier/test` folder, create a file called `lit.cfg.py` that contains the following content:

    ```
    import lit.formats


    config.name = 'JSMinifier Test'
    config.test_format = lit.formats.ShTest(True)
    config.suffixes = ['.js']
    ```

    Here, the snippet is providing LIT with some information. The `config` variable here is a Python object that will be populated later when this script is loaded into LIT's runtime. It's basically a registry with predefined fields that carry configuration values, along with custom fields that can be added by `lit.*.py` scripts at any time.

The config.test_format field suggests that LIT will run every test inside a shell environment (in the ShTest format), while the config.suffixes field suggests that only files with .js in their filename suffix will be treated as test cases (that is, all the JavaScript files).

2. Following on from the code snippet in the previous step, LIT now needs two other pieces of information: the *root path* to the test files and the *working directory*:

```
...
config.suffixes = ['.js']

config.test_source_root = os.path.dirname(__file__)
config.test_exec_root = os.path.join(config.my_obj_root,
'test')
```

For config.test_source_root, it's simply pointing to /JSMinifier/test. On the other hand, config.test_exec_root, which is the working directory, is pointing to a place whose parent folder is the value of a custom configuration field, my_obj_root. While it will be introduced later, simply put, it points to the build folder path. In other words, config.test_exec_root will eventually have a value of /JSMinifier/build/test.

3. The %jsm directive we saw earlier in test.js is used as a placeholder that will eventually be replaced with the real/absolute path of the js-minifier executable. The following lines will set up the replacements:

```
...
config.test_exec_root = os.path.join(config.my_obj_root,
'test')

config.substitutions.append(('%jsm',
    os.path.join(config.my_obj_root, 'js-minifier')))
```

This code adds a new entry to the config.substitutions field, which makes LIT replace every %jsm occurrence in the test files with the /JSMinifier/build/js-minifier value. This wraps up all the content in lit.cfg.py.

4. Now, create a new file called `lit.site.cfg.py.in` and put it under the `/JSMinifier/test` folder. The first part of this file looks like this:

```
import os
config.my_src_root = r'@CMAKE_SOURCE_DIR@'
config.my_obj_root = r'@CMAKE_BINARY_DIR@'
```

The mystery `config.my_obj_root` field is finally resolved here, but instead of pointing to a normal string, it is assigned to a weird value called `@CMAKE_BINARY_DIR@`. Again, this will be replaced by CMake with the real path later. The same goes for the `config.my_src_root` field.

5. Finally, `lit.site.cfg.py.in` is wrapped up by these lines:

```
...
lit_config.load_configure(
    config, os.path.join(config.my_src_root, 'test/
      lit.cfg.py'))
```

Even though this snippet is pretty simple, it's a little hard to understand. Simply put, this file will eventually be *materialized* into another file, with all the variables clamped by @ being resolved and copied into the `build` folder. From there, it will *call back* the `lit.cfg.py` script we saw in the earlier steps. This will be explained later in this section.

6. Finally, it's time to replace those weird @-clamped strings with real values using CMake's `configure_file` function. In `/JSMinifier/test/CMakeLists.txt`, add the following line somewhere inside the file:

```
configure_file(lit.site.cfg.py.in
               lit.site.cfg.py @ONLY)
```

The `configure_file` function will replace all the @-clamped string occurrences in the input file (`lit.site.cfg.py.in`, in this case) with their CMake variable counterparts in the current CMake context.

For example, let's say there is a file called `demo.txt.in` that contains the following content:

```
name = "@FOO@"
age = @AGE@
```

Now, let's use `configure_file` in `CMakeLists.txt`:

```
set(FOO "John Smith")
set(AGE 87)
configure_file(demo.txt.in
               demo.txt @ONLY)
```

Here, the aforementioned replacement will kick in and generate an output file, `demo.txt`, that contains the following content:

```
name = "John Smith"
age = 87
```

7. Back to the `lit.site.cfg.py.in` snippets, since `CMAKE_SOURCE_DIR` and `CMAKE_BINARY_DIR` are always available, they point to the root source folder and the `build` folder, respectively. The resulting `/JSMinifier/build/test/lit.site.cfg.py` will contain the following content:

```
import os
config.my_src_root = r'/absolute/path/to/JSMinifier'
config.my_obj_root = r'/absolute/path/to/JSMinifier/
build'

lit_config.load_config(
    config, os.path.join(config.my_src_root, 'test/
      lit.cfg.py'))
```

With that, we have learned how to write LIT configuration scripts for our example project. Now, it is time to explain some details of how LIT works internally, and why we need so many files (`lit.cfg.py`, `lit.site.cfg.py.in`, and `lit.site.cfg.py`).

LIT internals

Let's look at the following diagram, which illustrates the workflow of running LIT tests in the demo project we just created:

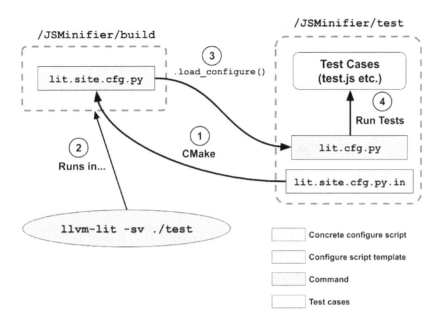

Figure 3.1 – The forking flow of LIT in our example project

Let's take a look at this diagram in more detail:

1. `lit.site.cfg.py.in` is copied to `/JSMinifier/build/lit.site.cfg.py`, which carries some CMake variable values.

2. The `llvm-lit` command is launched inside `/JSMinifier/build`. It will execute `lit.site.cfg.py` first.

3. `lit.site.cfg.py` then uses the `load_configure` Python function to load the main LIT configurations (`lit.cfg.py`) and run all the test cases.

The most crucial part of this diagram is explaining the roles of `lit.site.cfg.py` and `lit.site.cfg.py.in`: many parameters, such as the absolute path to the `build` folder, will remain unknown until the CMake configuration process is complete. So, a *trampoline* script – that is, `lit.site.cfg.py` – is placed inside the `build` folder to relay that information to the real test runner.

In this section, we learned how to write LIT configuration scripts for our out-of-tree example project. We also learned how LIT works under the hood. Knowing this can help you use LIT in a wide variety of projects, in addition to LLVM. In the next section, we will focus on `FileCheck`, a crucial and commonly used LIT utility that performs advanced pattern checking.

Learning useful FileCheck tricks

FileCheck is an advanced pattern checker that originates from LLVM. It has a similar role as the `grep` command-line tool available in Unix/Linux systems, but provides a more powerful yet straightforward syntax for line-based contexts. Furthermore, the fact that you can put `FileCheck` directives beside the testing targets makes the test cases self-contained and easy to understand.

Though basic `FileCheck` syntax is easy to get hands-on with, there are many other `FileCheck` functionalities that truly unleash the power of `FileCheck` and greatly improve your testing experiences – creating more concise testing scripts and parsing more complex program output, to name a few. This section will show you some of those skills.

Preparing for our example project

The `FileCheck` command-line tool needs to be built first. Similar to the previous section, building one of the `check-XXX` (phony) targets in the LLVM tree is the easiest way to do so. The following is an example of this:

```
$ ninja check-llvm-support
```

In this section, we are going to use an imaginary command-line tool called `js-obfuscator`, a JavaScript obfuscator, for our example. **Obfuscation** is a common technique that's used to hide intellectual properties or enforce security protections. For example, we could use a real-world JavaScript obfuscator on the following JavaScript code:

```
const onLoginPOST = (req, resp) => {
  if(req.name == 'admin')
    resp.send('OK');
  else
    resp.sendError(403);
}
myReset.post('/console', onLoginPOST);
```

This would transform it into the following code:

```
const t = "nikfmnsdzaO";
const aaa = (a, b) => {
  if(a.z[0] == t[9] && a.z[1] == t[7] &&…)
    b.f0(t[10] + t[2].toUpperCase());
  else
    b.f1(0x193);
```

```
}
G.f4(YYY, aaa);
```

This tool will try to make the original script as human-unreadable as possible. The challenge for the testing part is to verify its correctness while still reserving enough space for randomness. Simply put, `js-obfuscator` will only apply four obfuscation rules:

1. Only obfuscate local variable names, including formal parameters. The formal parameter names should always be obfuscated in *<lower case word><argument index number>* format. The local variable names will always be obfuscated into a combination of lowercase and uppercase letters.

2. If we are declaring functions with the arrow syntax – for example, `let foo = (arg1, arg2) => {...}` – the arrow and the left curly brace (`=> {`) need to be put in the next line.

3. Replace a literal number with the same value but in a different representation; for example, replacing *87* with *0x57* or *87.000*.

4. When you supply the tool with the `--shuffle-funcs` command-line option, shuffle the declaration/appearing order of the top-level functions.

Finally, the following JavaScript code is the example to be used with the `js-obfuscator` tool:

```
const square = x => x * x;
const cube = x => x * x * x;
const my_func1 = (input1, input2, input3) => {
    // TODO: Check if the arrow and curly brace are in the second
    // line
    // TODO: Check if local variable and parameter names are
    // obfuscated
    let intermediate = square(input3);
    let output = input1 + intermediate - input2;
    return output;
}
const my_func2 = (factor1, factor2) => {
    // TODO: Check if local variable and parameter names are
    // obfuscated
    let term2 = cube(factor1);
    // TODO: Check if literal numbers are obfuscated
    return my_func1(94,
```

```
              term2, factor2);
}
console.log(my_func2(1,2));
```

Writing FileCheck directives

The following steps are going to fill in all the TODO comments that appeared in the preceding code:

1. Going according to the line number, the first task is to check whether the local variables and parameters have been obfuscated properly. According to the spec, formal parameters have special renaming rules (that is, *<lower case word><argument index number>*), so using the normal CHECK directive with FileCheck's own regex syntax will be the most suitable solution here:

    ```
    // CHECK: my_func1 = ({{[a-z]+0}}, {{[a-z]+1}},
    // {{[a-z]+2}})
    const my_func1 = (input1, input2, input3) => {
        ...
    ```

    FileCheck uses a subset of regular expressions for pattern matching, which are enclosed by either {{...}} or [[...]] symbols. We will cover the latter one shortly.

2. This code looks pretty straightforward. However, the semantics of the code also need to be correct once obfuscation has been performed. So, in addition to checking the format, the succeeding references to those parameters need to be refactored as well, which is where FileCheck's pattern binding comes in:

    ```
    // CHECK: my_func1 = ([[A0:[a-z]+0]],
    // [[A1:[a-z]+1]], [[A2:[a-z]+2]])
    const my_func1 = (input1, input2, input3) => {
        // CHECK: square([[A2]])
        let intermediate = square(input3);
        ...
    ```

    This code binds the pattern of the formal parameters with the names A0 ~ A2 using the [[...]] syntax, in which the binding variable name and the pattern are divided by a colon: [[<binding variable>:<pattern>]]. On the reference sites of the binding variable, the same [[...]] syntax is used, but without the pattern part.

> **Note**
> A binding variable can have multiple definition sites. Its reference sites will read the last defined value.

3. Let's not forget the second rule – the arrow and left curly brace of the function header need to be put in the second line. To implement the concept of "the line after," we can use the CHECK-NEXT directive:

```
// CHECK: my_func1 = ([[A0:[a-z]+0]],
// [[A1:[a-z]+1]], [[A2:[a-z]+2]])
const my_func1 = (input1, input2, input3) => {
    // CHECK-NEXT: => {
```

Compared to the original CHECK directive, CHECK-NEXT will not only check if the pattern exists but also ensure that the pattern is in the line that follows the line matched by the previous directive.

4. Next, all the local variables and formal parameters are checked in my_func1:

```
// CHECK: my_func1 = ([[A0:[a-z]+0]],
// [[A1:[a-z]+1]], [[A2:[a-z]+2]])
const my_func1 = (input1, input2, input3) => {
    // CHECK: let [[IM:[a-zA-Z]+]] = square([[A2]]);
    let intermediate = square(input3);
    // CHECK: let [[OUT:[a-zA-Z]+]] =
    // CHECK-SAME: [[A0]] + [[IM]] - [[A1]];
    let output = input1 + intermediate - input2;
    // CHECK: return [[OUT]];
    return output;
}
```

As highlighted in the preceding code, the CHECK-SAME directive was used to match the succeeding pattern in the exact same line. The rationale behind this is that FileCheck expected different CHECK directives to be matched in different *lines*. So, let's say part of the snippet was written like this:

```
// CHECK: let [[OUT:[a-zA-Z]+]] =
// CHECK: [[A0]] + [[IM]] - [[A1]];
```

It will *only* match code that spread across two lines or more, as shown here:

```
let BGHr =
    r0 + jkF + r1;
```

It will throw an error otherwise. This directive is especially useful if you wish to avoid writing a super long line of checking statements, thus making the testing scripts more concise and readable.

5. Going into my_func2, now, it's time to check if the literal numbers have been obfuscated properly. The checking statement here is designed to accept any instances/patterns *except* the original numbers. Therefore, the CHECK-NOT directive will be sufficient here:

```
...
// CHECK: return my_func1(
// CHECK-NOT: 94
 return my_func1(94,
                  term2, factor2);
```

> **Note**
>
> The first CHECK directive is required. This is because CHECK-NOT will not move the cursor from the line before return my_func1(94. Here, CHECK-NOT will give a false negative without a CHECK directive to move the cursor to the correct line first.

In addition, CHECK-NOT is pretty useful to express the concept of *not <a specific pattern>…but <the correct pattern>* when it's used with CHECK-SAME, as we mentioned earlier.

For example, if the obfuscation rule states that all the literal numbers need to be obfuscated into their hexadecimal counterparts, then you can express the assertion of *don't want to see 94… but want to see 0x5E/0x5e at the same place instead* using the following code:

```
...
// CHECK: return my_func1
// CHECK-NOT: 94,
// CHECK-SAME: {{0x5[eE]}}
 return my_func1(94,
                  term2, factor2);
```

6. Now, only one obfuscation rule needs to be verified: when the `js-obfuscator` tool is supplied with an additional command-line option, `--shuffle-funcs`, which effectively shuffles all top-level functions, we need to check whether the top-level functions maintain certain ordering, even after they have been shuffled. In JavaScript, functions are resolved when they're called. This means that `cube`, `square`, `my_func1`, and `my_func2` can have an arbitrary ordering, as long as they're placed before the `console.log (...)` statement. To express this kind of flexibility, the `CHECK-DAG` directive can be pretty useful.

Adjacent `CHECK-DAG` directives will match texts in arbitrary orders. For example, let's say we have the following directives:

```
// CHECK-DAG: 123
// CHECK-DAG: 456
```

These directives will match the following content:

```
123
456
```

They will also match the following content:

```
456
123
```

However, this freedom of ordering will not hold across either a `CHECK` or `CHECK-NOT` directive. For example, let's say we have these directives:

```
// CHECK-DAG: 123
// CHECK-DAG: 456
// CHECK: 789
// CHECK-DAG: abc
// CHECK-DAG: def
```

These directives will match the following text:

```
456
123
789
def
abc
```

However, they will not match the following text:

```
456
789
123
def
abc
```

7. Back to our motivated example, the obfuscation rule can be checked by using the following code:

```
...
// CHECK-DAG: const square =
// CHECK-DAG: const cube =
// CHECK-DAG: const my_func1 =
// CHECK-DAG: const my_func2 =
// CHECK: console.log
console.log(my_func2(1,2));
```

However, function shuffling will only happen if an additional command-line option is supplied to the tool. Fortunately, FileCheck provides a way to multiplex different *check suites* into a single file, where each suite can define how it runs and separates the checks from other suites.

8. The idea of the check prefix in FileCheck is pretty simple: you can create a *check suite* that runs independently with other suites. Instead of using the CHECK string, each suite will replace it with another string in all the directives mentioned earlier (CHECK-NOT and CHECK-SAME, to name a few), including CHECK itself, in order to distinguish it from other suites in the same file. For example, you can create a suite with the YOLO prefix so that that part of the example now looks as follows:

```
// YOLO: my_func2 = ([[A0:[a-z]+0]], [[A1:[a-z]+1]])
const my_func2 = (factor1, factor2) => {
...
// YOLO-NOT: return my_func1(94,
// YOLO-SAME: return my_func1({{0x5[eE]}},
return my_func1(94,
                term2, factor2);
...
```

To use a custom prefix, it needs to be specified in the `--check-prefix` command-line option. Here, the `FileCheck` command invocation will look like this:

```
$ cat test.out.js | FileCheck --check-prefix=YOLO test.js
```

9. Finally, let's go back to our example. The last obfuscation rule can be solved by using an alternative prefix for those `CHECK-DAG` directives:

```
...
// CHECK-SHUFFLE-DAG: const square =
// CHECK-SHUFFLE-DAG: const cube =
// CHECK-SHUFFLE-DAG: const my_func1 =
// CHECK-SHUFFLE-DAG: const my_func2 =
// CHECK-SHUFFLE: console.log
console.log(my_func2(1,2));
```

This must be combined with the default check suite. All the checks mentioned in this section can be run in two separate commands, as follows:

```
# Running the default check suite
$ js-obfuscator test.js | FileCheck test.js
# Running check suite for the function shuffling option
$ js-obfuscator --shuffle-funcs test.js | \
  FileCheck --check-prefix=CHECK-SHUFFLE test.js
```

In this section, we have shown some advanced and useful `FileCheck` skills through our example project. These skills provide you with different ways to write validation patterns and make your LIT test script more concise.

So far, we have been talking about the testing methodology, which runs tests in a shell-like environment (that is, in the `ShTest` LIT format). In the next section, we are going to introduce an alternative LIT framework – the TestSuite framework and testing format that was originated from the `llvm-test-suite` project – which provides a *different kind* of useful testing methodology for LIT.

Exploring the TestSuite framework

In the previous sections, we learned how regression tests were performed in LLVM. More specifically, we looked at the `ShTest` testing format (recalling the `config.test_format = lit.formats.ShTest(...)` line), which basically runs end-to-end tests in a shell script fashion. The `ShTest` format provides more flexibility when it comes to validating results since it can use the `FileCheck` tool we introduced in the previous section, for example.

This section is going to introduce another kind of testing format: **TestSuite**. The TestSuite format is part of the `llvm-test-suite` project – a collection of test suites and benchmarks created for testing and benchmarking LLVM. Similar to `ShTest`, this LIT format is also designed to run end-to-end tests. However, TestSuite aims to make developers' lives easier when they want to integrate *existing* executable-based test suites or benchmark codebases. For example, if you want to use the famous **SPEC benchmark** as one of your test suites, all you need to do is add a build description and the expected output in plain text. This is also useful when your testing logic cannot be expressed using a **textual testing script**, as we saw in previous sections.

In this section, we will learn how to import an existing test suite or benchmark codebase into the `llvm-test-suite` project.

Preparing for our example project

First, please follow the instructions at the beginning of this chapter to build `llvm-test-suite`.

The rest of the section is going to use a pseudo test suite project called **GeoDistance**. The GeoDistance project uses C++ and a GNU `Makefile` to build a command-line tool, `geo-distance`, that calculates and prints out the total distance of a path constructed by a list of latitude and longitude pairs provided by the input file.

It should have the following folder structure:

```
GeoDistance
    |___ helper.cpp
    |___ main.cpp
    |___ sample_input.txt
    |___ Makefile
```

Here, the `Makefile` looks like this:

```
FLAGS := -DSMALL_INPUT -ffast-math
EXE := geo-distance
OBJS := helper.o main.o

%.o: %.cpp
    $(CXX) $(FLAGS) -c $^
$(EXE): $(OBJS)
    $(CXX) $(FLAGS) $< -o $@
```

To run the `geo-distance` command-line tool, use the following command:

```
$ geo-distance ./sample_input.txt
```

This prints out the floating-point distance to `stdout`:

```
$ geo-distance ./sample_input.txt
94.873467
```

The floating-point precision requirement here is `0.001`.

Importing code into llvm-test-suite

Basically, there are only two things we need to do to import existing test suites or benchmarks into `llvm-test-suite`:

- Use CMake as the build system
- Compose verification rules

To use CMake as the build system, the project folder needs to be put under the `MultiSource/Applications` subdirectory inside the `llvm-test-suite` source tree. Then, we need to update the enclosing `CMakeLists.txt` accordingly:

```
# Inside MultiSource/Applications/CMakeLists.txt
...
add_subdirectory(GeoDistance)
```

To migrate from our GNU `Makefile` to `CMakeLists.txt`, instead of rewriting it using the built-in CMake directives such as `add_executable`, LLVM provides some handy functions and macros for you:

```
# Inside MultiSource/Applications/GeoDistance/CMakeLists.txt
# (Unfinished)
llvm_multisource(geo-distance)
llvm_test_data(geo-distance sample_input.txt)
```

There are some new CMake directives here. `llvm_multisource` and its sibling, `llvm_singlesource`, add a new executable build target from multiple source files or only a single source file, respectively. They're basically `add_executable`, but as shown in the previous code, you can choose to leave the source file list empty, and it will use all the C/C++ source files shown in the current directory as input.

> **Note**
>
> If there are multiple source files but you're using `llvm_singlesource`, every source file will be treated as a standalone executable.

`llvm_test_data` copies any resource/data files you want to use during runtime to the proper working directory. In this case, it's the `sample_input.txt` file.

Now that the skeleton has been set up, it's time to configure the compilation flags using the following code:

```
# Inside MultiSource/Applications/GeoDistance/CMakeLists.txt
# (Continue)
list(APPEND CPPFLAGS -DSMALL_INPUT)
list(APPEND CFLAGS -ffast-math)


llvm_multisource(geo-distance)
llvm_test_data(geo-distance sample_input.txt)
```

Finally, TestSuite needs to know how to run the test and how to verify the result:

```
# Inside MultiSource/Applications/GeoDistance/CMakeLists.txt
# (Continue)
…
set(RUN_OPTIONS sample_input.txt)
set(FP_TOLERANCE 0.001)
llvm_multisource(geo-distance)
…
```

The RUN_OPTIONS CMake variable is pretty straightforward – it provides the command-line options for the testing executable.

For the verification part, by default, TestSuite will use an enhanced diff to compare the output of stdout and the exit code against files whose filename end with .reference_output.

For example, in our case, a GeoDistance/geo-distance.reference_output file is created with the expected answer and exit status code:

```
94.873
exit 0
```

You might find that the expected answer here is slightly different from the output at the beginning of this section (94.873467), and that's because the comparison tool allows you to designate the desired floating-point precision, which is controlled by the FP_TOLERANCE CMake variable shown previously.

In this section, we learned how to leverage the llvm-test-suite project and its TestSuite framework to test executables that are either from an existing codebase or are unable to express testing logic using textual scripts. This will help you become more efficient in testing different kinds of projects using LIT.

Summary

LIT is a general-purpose testing framework that can not only be used inside LLVM, but also arbitrary projects with little effort. This chapter tried to prove this point by showing you how to integrate LIT into an out-of-tree project without even needing to build LLVM. Second, we saw FileCheck – a powerful pattern checker that's used by many LIT test scripts. These skills can reinforce the expressiveness of your testing scripts. Finally, we presented you with the TestSuite framework, which is suitable for testing different kinds of program and complements the default LIT testing format.

In the next chapter, we will explore another supporting framework in the LLVM project: **TableGen**. We will show you that TableGen is also a *general toolbox* that can solve problems in out-of-tree projects, albeit almost being exclusively used by backend development in LLVM nowadays.

Further reading

Currently, the source code for FileCheck – written in C++ – is still inside LLVM's source tree. Try to replicate its functionality using Python (`https://github.com/mull-project/FileCheck.py`), which will effectively help you use FileCheck without building LLVM, just like LIT!

4
TableGen Development

TableGen is a **domain-specific language** (**DSL**) that was originally developed in **Low-Level Virtual Machine** (**LLVM**) to express processors' **instruction set architecture** (**ISA**) and other hardware-specific details, similar to the **GNU Compiler Collection's** (**GCC's**) **Machine Description** (**MD**). Thus, many people learn TableGen when they're dealing with LLVM's backend development. However, TableGen is not just for describing hardware specifications: it is a *general DSL* useful for any tasks that involve non-trivial *static and structural data*. LLVM has also been using TableGen on parts outside the backend. For example, Clang has been using TableGen for its command-line options management. People in the community are also exploring the possibility to implement **InstCombine** rules (LLVM's **peephole optimizations**) in TableGen syntax.

Despite TableGen's universality, the language's core syntax has never been widely understood by many new developers in this field, creating lots of copy-and-pasted boilerplate TableGen code in LLVM's code base since they're not familiar with the language *itself*. This chapter tries to shed a little bit of light on this situation and show the way to apply this amazing technique to a wide range of applications.

The chapter starts with an introduction to common and important TableGen syntax, which prepares you for writing a delicious donut recipe in TableGen as a practice, culminating in a demonstration of TableGen's universality in the second part. Finally, the chapter will end with a tutorial to develop a custom *emitter*, or a **TableGen backend**, to convert those nerdy sentences in the TableGen recipe into normal plaintext descriptions that can be put in the kitchen.

Here is the list of the sections we will be covering:

- Introduction to TableGen syntax
- Writing a donut recipe in TableGen
- Printing a recipe via the TableGen backend

Technical requirements

This chapter focuses on one tool in the `utils` folder: `llvm-tblgen`. To build it, launch the following command:

```
$ ninja llvm-tblgen
```

> **Note**
>
> If you chose to build `llvm-tblgen` in **Release** mode regardless of the global build type, using the `LLVM_OPTIMIZED_TABLEGEN` CMake variable introduced in the first chapter, you might want to change that setting since it's always better to have a debug version of `llvm-tblgen` in this chapter.

All of the source code in this chapter can be found in this GitHub repository: `https://github.com/PacktPublishing/LLVM-Techniques-Tips-and-Best-Practices-Clang-and-Middle-End-Libraries/tree/main/Chapter04`.

Introduction to TableGen syntax

This section serves as a quick tour of all the important and common TableGen syntax, providing all the essential knowledge to get hands-on, writing a donut recipe in TableGen in the next section.

TableGen is a domain-specific programming language used for modeling custom data layouts. Despite being a programming language, it does something quite different from conventional languages. **Conventional programming languages** usually describe *actions* performed on the (input) data, how they interact with the environment, and how they generate results, regardless of the programming paradigms (imperative, functional, event-driven…) you adopt. TableGen, in contrast, barely describes any actions.

TableGen is designed only to describe structural **static data**. First, developers define the layout—which is essentially just a table with many fields—of their desired data structure. They then need to fill data into those layouts *right away* as most of the fields are populated/initialized. The latter part is probably what makes TableGen unique: many programming languages or frameworks provide ways to design your domain-specific data structures (for example, Google's **Protocol Buffers**), but in those scenarios, data is usually filled in **dynamically**, mostly in the code that consumes the DSL part.

Structured Query Language (**SQL**) shares many aspects with TableGen: both SQL and TableGen (only) handle structural data and have a way to define the layout. In SQL, it's TABLE; and in TableGen, it's class, which will be introduced later on in this section. However, SQL provides much more functions other than crafting the layout. It can also query (actually, that's where its name came from: **Structured Query Language**) and update data dynamically, which are absent in TableGen. However, later in this chapter, you will see that TableGen provides a nice framework to flexibly process and *interpret* this TableGen-defined data.

We'll now introduce four important TableGen constructions, as follows:

- Layout and records
- Bang operators
- Multiclass
- The **Directed-Acyclic Graph** (**DAG**) data type

Layout and records

Given the fact that TableGen is just a more fancy and expressive way to describe structural data, it's pretty straightforward to think that there is a primitive representation for the data's **layout**, and representation for the *instantiated* data. The layout is realized by the `class` syntax, as shown in the following code snippet:

```
class Person {
  string Name = "John Smith";
  int Age;
}
```

As shown here, a class is similar to a struct in C and many other programming languages, which only contains a group of data fields. Each field has a type, which can be any of the primitive types (`int`, `string`, `bit`, and so on) or another user-defined `class` type. A field can also assign a default value such as `John Smith`.

After looking a layout, it's time to create an instance (or a **record**, in TableGen's terms), out of it, as follows:

```
def john_smith : Person;
```

Here, `john_smith` is a record using `Person` as a template so that it also has two fields—Name and Age—with the `Name` field filled with the value `John Smith`. This looks pretty straightforward, but recall that TableGen should define static data and that *most* fields should be filled with values. Also, in this case, the `Age` field is still left uninitialized. You can populate its value by *overriding* with a bracket closure and statements within, as follows:

```
def john_smith : Person {
  let Age = 87;
}
```

You can even define new fields specifically for the `john_smith` record, as follows:

```
def john_smith : Person {
  let Age = 87;
  string Job = "Teacher";
}
```

Just be aware that you can only override fields (using the `let` keyword) that have been declared, just as with many other programming languages.

Bang operators

Bang operators are a group of functions performing simple tasks such as basic arithmetic or casting on values in TableGen. Here is a simple example of converting kilograms to grams:

```
class Weight<int kilogram> {
  int Gram = !mul(kilogram, 1000);
}
```

Common operators include arithmetic and bitwise operators (to name but a few), and some of these are outlined here:

- `!add(a, b)`: For arithmetic addition
- `!sub(a, b)`: For arithmetic subtraction
- `!mul(a, b)`: For arithmetic multiplication
- `!and(a, b)`: For logical AND operations
- `!or(a, b)`: For logical OR operations
- `!xor(a, b)`: For logical XOR operations

We also use conditional operators, and a few are outlined here:

- `!ge(a, b)`: Returns 1 if a `>=` b, and 0 otherwise
- `!gt(a, b)`: Returns 1 if a `>` b, and 0 otherwise
- `!le(a, b)`: Returns 1 if a `<=` b, and 0 otherwise
- `!lt(a, b)`: Returns 1 if a `<` b, and 0 otherwise
- `!eq(a, b)`: Returns 1 if a `==` b, and 0 otherwise

Other interesting operators include the following:

- `!cast<type>(x)`: This operator performs type casting on the x operand, according to the `type` parameter. In cases where the type is a numerical type, such as with `int` or `bits`, this performs normal arithmetic type casting. In some special cases, we have the following scenarios:

  If `type` is string and x is a record, this returns the record's name.

  If x is a string, it is treated as the name of a record. TableGen will look up all the record definitions so far and cast the one with the name of x and return it with a type that matches the `type` parameter.

- !if(pred, then, else): This operator returns the then expression if pred is 1, and returns the else expression otherwise.
- !cond(cond1 : val1, cond2 : val2, ..., condN : valN): This operator is an enhanced version of the !if operator. It will continuously evaluate cond1...condN until one of the expressions returns 1, before returning its associated val expression.

> **Note**
>
> Unlike functions, which are evaluated during runtime, bang operators are more like *macros*, which are evaluated during build time—or in TableGen's terminology, when those syntaxes are processed by TableGen backends.

Multiclass

There are many cases where we want to define multiple records at once. For example, the following snippet tries to create *auto part* records for multiple cars:

```
class AutoPart<int quantity> {...}


def car1_fuel_tank : AutoPart<1>;
def car1_engine : AutoPart<1>;
def car1_wheels : AutoPart<4>;
...
def car2_fuel_tank : AutoPart<1>;
def car2_engine : AutoPart<1>;
def car2_wheels : AutoPart<4>;
...
```

We can further simplify these by using the multiclass syntax, as follows:

```
class AutoPart<int quantity> {...}

multiclass Car<int quantity> {
  def _fuel_tank : AutoPart<quantity>;
  def _engine : AutoPart<quantity>;
  def _wheels : AutoPart<!mul(quantity, 4)>;
  ...
}
```

When creating record instances, use the `defm` syntax instead of `def`, as follows:

```
defm car1 : Car<1>;
defm car2 : Car<1>;
```

Thus, at the end, it will still generate records with names such as `car1_fuel_tank`, `car1_engine`, `car2_fuel_tank`, and so on.

Despite having `class` in its name, `multiclass` has nothing to do with a class. Instead of describing the layout of a record, `multiclass` acts as a template to *generate* records. Inside a `multiclass` template are the prospective records to be created and the records' name *suffix* after the template is expanded. For example, the `defm car1 : Car<1>` directive in the preceding snippet will eventually be expanded into three `def` directives, as follows:

- `def car1_fuel_tank : AutoPart<1>;`
- `def car1_engine : AutoPart<1>;`
- `def car1_wheels : AutoPart<!mul(1, 4)>;`

As you can see in the preceding list, the name suffixes we found inside `multiclass` (for instance, `_fuel_tank`) was concatenated with the name appearing after `defm`— `car1` in this case. Also, the `quantity` template argument from `multiclass`, was also instantiated into every expanded record.

In short, `multiclass` tries to extract common parameters from multiple record instances and make it possible to create them at once.

The DAG data type

In addition to conventional data types, TableGen has a pretty unique first-class type: the `dag` type that is used for expressing DAG instances. To create a DAG instance, you can use the following syntax:

```
(operator operand1, operand2,…, operandN)
```

While the `operator` can only be a record instance, operands (`operand1...operandN`) can have arbitrary types. Here is an example of trying to model an arithmetic expression, `x * 2 + y + 8 * z`:

```
class Variable {…}
class Operator {…}
class Expression<dag expr> {…}

// define variables
def x : Variable;
def y : Variable;
def z : Variable;

// define operators
def mul : Operator;
def plus : Operator;

// define expression
def tmp1 : Expression<(mul x, 2)>;
def tmp2 : Expression<(mul 8, z)>;
def result : Expression<(plus tmp1, tmp2, y)>;
```

Optionally, you can associate `operator` and/or each operand with a *tag*, as follows:

```
…
def tmp1 : Expression<(mul:$op x, 2)>;
def tmp2 : Expression<(mul:$op 8, z)>;
def result : Expression<(plus tmp1:$term1, tmp2:$term2,
y:$term3)>;
```

A tag always starts with a dollar sign, $, followed by a user-defined tag name. These tags provide a *logical* description of each `dag` component and can be useful when processing DAGs in the TableGen backend.

In this section, we have gone through the principal components of the TableGen language and introduced some essential syntax. In the next section, we are going to get hands-on, writing a delicious donut recipe using TableGen.

Writing a donut recipe in TableGen

With the knowledge from previous sections, it's time to write our own donut recipe! We'll proceed as follows:

1. The first file to create is `Kitchen.td`. It defines the environment for cooking, including measuring units, equipment, and procedures, to name but a few aspects. We are going to start with the measuring units, as follows:

```
class Unit {
  string Text;
  bit Imperial;
}
```

Here, the `Text` field is the textual format showing on the recipe, and `Imperial` is just a Boolean flag marking whether this unit is imperial or metric. Each weight or volume unit will be a record inheriting from this class—have a look at the following code snippet for an example of this:

```
def gram_unit : Unit {
  let Imperial = false;
  let Text = "g";
}
def tbsp_unit : Unit {
  let Imperial = true;
  let Text = "tbsp";
}
```

There are plenty of measuring units we want to create, but the code is already pretty lengthy. A way to simplify and make it more readable is by using `class` template arguments, as follows:

```
class Unit<bit imperial, string text> {
  string Text = text;
  bit Imperial = imperial;
}
def gram_unit : Unit<false, "g">;
def tbsp_unit : Unit<true, "tbsp">;
```

In contrast to C++'s template arguments, the template arguments in TableGen only accept concrete values. They're just an alternative way to assign values to fields.

2. Since TableGen doesn't support floating-point numbers, we need to define some way to express numberings, such as **1 and ¼ cups** or **94.87g of flour**. One solution is to use a *fixed point*, as follows:

```
class FixedPoint<int integral, int decimal = 0> {
  int Integral = integral;
  int DecimalPoint = decimal;
}
def one_plus_one_quarter : FixedPoint<125, 2>; // Shown
as 1.25
```

With the `Integral` and `DecimalPoint` fields mentioned, the value represented by this `FixedPoint` class is equal to the following formula:

*Integral * 10^(-DecimalPoint)*

Since ¼, ½, and ¾ are apparently commonly used in measuring (especially for imperial units such as a US cup), it's probably a good idea to use a helper class to create them, as follows:

```
class NplusQuarter<int n, bits<2> num_quarter> :
FixedPoint<?, 2> {…}
def one_plus_one_quarter : NplusQuarter<1,1>; // Shown as
1.25
```

This will make expressing quantities such as N and ¼ cups or N and ½ cups a lot easier.

TableGen classes also have inheritance—a class can inherit one or more classes. Since TableGen doesn't have the concept of member functions/methods, inheriting `class` is simply just integrating its fields.

3. To implement `NplusQuarter`, especially the conversion from the `NplusQuarter` class template parameters to that of `FixedPoint`, we need some simple arithmetic calculations, which is where TableGen's bang operators come into place, as follows:

```
class NplusQuarter<int n, bits<2> num_quarter> :
FixedPoint<?, 2> {
  int Part1 = !mul(n, 100);
  int Part2 = !mul(25, !cast<int>(num_quarter{1...0}));
  let Integral = !add(Part1, Part2);
}
```

Another interesting syntax that appeared is the *bit extraction (or slicing)* on the num_quarter variable. By writing num_quarter{1…0}, this gives you a bits value that is equal to the 0th and first bit of num_quarter. There are some other variants of this technique. For example, it can slice a non-continuous range of bits, as follows:

```
num_quarter{8…6,4,2…0}
```

Or, it can extract bits in reversed ordering, as follows:

```
num_quarter{1…7}
```

> **Note**
> You might wonder why the code needs to extract the smallest 2 bits *explicitly* even it has declared that num_quarter has a width of 2 bits (the bits<2> type). It turned out that for some reason, TableGen will not stop anyone from assigning values greater than 3 into num_quarter, like this: def x : NplusQuarter<1,999>.

4. With the measuring units and number format, we can finally deal with the ingredients needed for this recipe. First, let's use a separated file, Ingredients.td, to store all the ingredient records. To use all the things mentioned earlier, we can import Kitchen.td by using the include syntax, as follows:

    ```
    // In Ingredients.td…
    include "Kitchen.td"
    ```

    Then, a base class of all ingredients is created to carry some common fields, as follows:

    ```
    class IngredientBase<Unit unit> {
      Unit TheUnit = unit;
      FixedPoint Quantity = FixedPoint<0>;
    }
    ```

Each kind of ingredient is represented by a class derived from `IngredientBase`, with parameters to specify the quantity needed by a recipe, and the unit used to measure this ingredient. Take milk, for example, as shown in the following code snippet:

```
class Milk<int integral, int num_quarter> :
IngredientBase<cup_unit> {
  let Quantity = NplusQuarter<integral, num_quarter>;
}
```

The `cup_unit` put at the template argument for `IngredientBase` tells us that milk is measured by a US cup unit, and its quantity is to be determined later by the `Milk` class template arguments.

When writing a recipe, each required ingredient is represented by a record created from one of these ingredient `class` types:

```
def ingredient_milk : Milk<1,2>; // Need 1.5 cup of milk
```

5. Some ingredients, however, always come together—for example, lemon peel and lemon juice, egg yolk, and egg white. That is, if you have two egg yolks, then there must be two servings of egg white. However, if we need to create a record and assign a quantity for each of the ingredients one by one, there will be a lot of duplicate code. A more elegant way to solve this problem is by using TableGen's `multiclass` syntax.

Taking the following egg example, assume we want to create `WholeEgg`, `EggWhite`, and `EggYolk` records at once with the same quantity, and define the `multiclass` first:

```
multiclass Egg<int num> {
  def _whole : WholeEgg {
    let Quantity = FixedPoint<num>;
  }
  def _yolk : EggYolk {
    let Quantity = FixedPoint<num>;
  }
  def _white : EggWhite {
    let Quantity = FixedPoint<num>;
  }
}
```

When writing the recipe, use the `defm` syntax to create `multiclass` records, as follows:

```
defm egg_ingredient : Egg<3>;
```

After using `defm`, three records will actually be created: `egg_ingredient_whole`, `egg_ingredient_yolk`, and `egg_ingredient_white`, inheriting from `WholeEgg`, `EggYolk`, and `EggWhite`, respectively.

6. Finally, we need a way to describe the steps to make a donut. Many recipes have some preparation steps that don't need to be done in a specific order. Take the donut recipe here, for example: preheating the oil can be done at any time before the donuts are ready to be fried. Thus, it might be a good idea to express baking steps in a `dag` type.

   Let's first create the `class` to represent a baking step, as follows:

```
class Step<dag action, Duration duration, string custom_
format> {
    dag Action = action;
    Duration TheDuration = duration;
    string CustomFormat = custom_format;
    string Note;
}
```

The `Action` field carries the baking instructions and information about the ingredients used. Here is an example:

```
def mix : Action<"mix",…>;
def milk : Milk<…>;
def flour : Flour<…>;
def step_mixing : Step<(mix milk, flour), …>;
```

`Action` is just a class used for describing movements. The following snippet represents the fact that `step_mixing2` is using the outcome from `step_mixing` (maybe a raw dough) and mixing it with butter:

```
…
def step_mixing : Step<(mix milk, flour), …>;
def step_mixing2 : Step<(mix step_mixing, butter), …>;
```

Eventually, all of the `Step` records will form a DAG, in which a vertex will either be a `step` or an ingredient record.

We're also annotating our `dag` operator and operand with tags, as follows:

```
def step_mixing2 : Step<(mix:$action step_mixing:$dough,
butter)>
```

In the previous section, *Introduction to TableGen syntax*, we said that these `dag` tags have no immediate effect in TableGen code, except affecting how TableGen backends handle the current record—for example, if we have a `string` type field, `CustomFormat`, in the `Step` class, as follows:

```
def step_prep : Step<(heat:$action fry_oil:$oil, oil_
temp:$temp)> {
   let CustomFormat = "$action the $oil until $temp";
}
```

With the field content shown, we can replace `$action`, `$oil`, and `$temp` in the string with the textual representation of those records, generating a string such as *heat the peanut oil until it reaches 300 F*.

And that wraps up this section of this chapter. In the next section, the goal is to develop a custom TableGen backend to take the TableGen version recipe here as input and print out a normal plaintext recipe.

Printing a recipe via the TableGen backend

Following up on the last part of the previous section, after composing the donut recipe in TableGen's syntax, it's time to print out a *normal* recipe from that via a custom-built TableGen backend.

> **Note**
>
> Please don't confuse a **TableGen backend** with a **LLVM backend**: the former converts (or transpiles) TableGen files into an *arbitrary textual content*, C/C++ header files being the most common form. An LLVM backend, on the other hand, lowers LLVM **intermediate representations** (**IR**) into low-level assembly code.

In this section, we're developing the TableGen backend to print the donut we composed in the previous section into content, like this:

```
=======Ingredients=======
1. oil 500 ml
2. flour 300 g
3. milk 1.25 cup
4. whole egg 1
5. yeast 1.50 tsp
6. butter 3.50 tbsp
7. sugar 2.0 tbsp
8. salt 0.50 tsp
9. vanilla extract 1.0 tsp


=======Instructions=======
1. use deep fryer to heat oil until 160 C
2. use mixer to mix flour, milk, whole egg, yeast, butter,
sugar, salt, and vanilla extract. stir in low speed.
3. use mixer to mix outcome from (step 2). stir in medium
speed.
4. use bowl to ferment outcome from (step 3).
5. use rolling pin to flatten outcome from (step 4).
6. use cutter to cut outcome from (step 5).
7. use deep fryer to fry outcome from (step 1) and outcome from
(step 6).
```

First, we will give an overview of `llvm-tblgen`, the program for driving the TableGen translation process. Then, we will show you how to develop our recipe-printing TableGen backend. Finally, we'll show you how to integrate our backend into the `llvm-tblgen` executable.

TableGen's high-level workflow

The TableGen backend takes in-memory representation (in the form of C++ objects) of the TableGen code we just learned and transforms it into arbitrary **textual content**. The whole process is driven by the llvm-tblgen executable, whose workflow can be illustrated by this diagram:

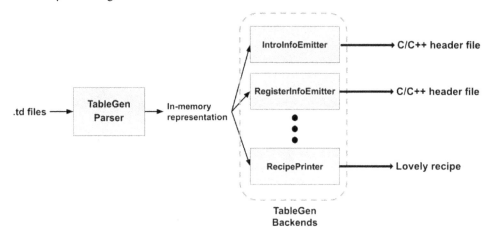

Figure 4.1 – Workflow of llvm-tblgen

TableGen code's in-memory representation (which consists of C++ types and APIs) plays an important role in the TableGen backend development. Similar to LLVM IR, it is organized *hierarchically*. Starting from the top level, here is a list of its hierarchy, where each of the items is a C++ class:

1. RecordKeeper: A collection (and owner) of all Record objects in the current translation unit.

2. Record: Represents a record or a class. The enclosing fields are represented by RecordVal. If it's a class, you can also access its template arguments.

3. RecordVal: Represents a *pair* of record fields and their initialized value, along with supplementary information such as the field's type and source location.

4. Init: Represents the initialized value of a field. It is a parent class of many, which represents different types of initialized values—For example, IntInit for integer values and DagInit for DAG values.

To give you a little task on the practical aspect of a TableGen backend, here is the skeleton of it:

```
class SampleEmitter {
  RecordKeeper &Records;
public:
  SampleEmitter(RecordKeeper &RK) : Records(RK) {}
  void run(raw_ostream &OS);
};
```

This emitter basically takes a `RecordKeeper` object (passed in by the constructor) as the input and prints the output into the `raw_ostream` stream—the function argument of `SampleEmitter::run`.

In the next section, we're going to show you how to set up the development environment and get hands- on, writing a TableGen backend.

Writing the TableGen backend

In this section, we're showing you the steps of writing a backend to print out recipes written in TableGen. Let's start with the setup.

Project setup

To get started, LLVM has already provided a skeleton for writing a TableGen backend. So, please copy the `llvm/lib/TableGen/TableGenBackendSkeleton.cpp` file from the LLVM Project's source tree into the `llvm/utils/TableGen` folder, as follows:

```
$ cd llvm
$ cp lib/TableGen/TableGenBackendSkeleton.cpp \
     utils/TableGen/RecipePrinter.cpp
```

Then, refactor the `cSkeletonEmitter` class into `RecipePrinter`.

`RecipePrinter` has the following workflow:

1. Collect all baking steps and ingredient records.
2. Print individual ingredients in textual formats using individual functions to print measuring units, temperature, equipment, and so on in textual formats.
3. Linearize the DAG of all baking steps.
4. Print each linearized baking step using a function to print custom formatting.

We're not going to cover all the implementation details since lots of backend codes are actually not directly related to TableGen (text formatting and string processing, for example). Therefore, the following subsections only focus on how to retrieve information from TableGen's in-memory objects.

Getting all the baking steps

In the TableGen backend, a TableGen record is represented by the Record C++ class. When we want to retrieve all the records derived from a specific TableGen class, we can use one of the functions of RecordKeeper: getAllDerivedDefinitions. For instance, let's say we want to fetch all the baking steps records that derived from the Step TableGen class in this case. Here is how we do with getAllDerivedDefinitions:

```
// In RecipePrinter::run method...
std::vector<Record*> Steps = Records.
getAllDerivedDefinitions("Step");
```

This gives us a list of Record pointers that represent all of the Step records.

> **Note**
>
> For the rest of this section, we will use Record in this format (with Courier font face) to refer to the C++ counterpart of a TableGen record.

Retrieving field values

Retrieving field values from Record is probably the most basic operation. Let's say we're working on a method for printing Unit record objects introduced earlier, as follows:

```
void RecipePrinter::printUnit(raw_ostream& OS, Record*
UnitRecord) {
   OS << UnitRecord->getValueAsString("Text");
}
```

The Record class provides some handy functions, such as getValueAsString, to retrieve the value of a field and try to convert it into a specific type so that you don't need to retrieve the RecordVal value of a specific field (in this case, the Text field) before getting the real value. Similar functions include the following:

- Record* getValueAsDef(StringRef FieldName)
- bool getValueAsBit(StringRef FieldName)

- `int64_t getValueAsInt(StringRef FieldName)`
- `DagInit* getValueAsDag(StringRef FieldName)`

In addition to these utility functions, we sometimes just want to check if a specific field exists in a record. In such cases, call `Record::getValue(StringRef FieldName)` and check if the returned value is null. But just be aware that not every field *needs* to be initialized; you may still need to check if a field exists, but is uninitialized. When that happens, let `Record::isValueUnset` help you.

> **Note**
>
> TableGen actually uses a special `Init` class, `UnsetInit`, to represent an uninitialized value.

Type conversion

`Init` represents initialization values, but most of the time we're not directly working with it but with one of its children's classes.

For example, `StepOrIngredient` is an `Init` type object that represents either a `Step` record or an ingredient record. It would be easier for us to convert it to its underlying `DefInit` object since `DefInit` provides richer functionalities. We can use the following code to typecast the `Init` type `StepOrIngredient` into a `DefInit` type object:

```
const auto* SIDef = cast<const DefInit>(StepOrIngredient);
```

You can also use `isa<...>(...)` to check its underlying type first, or `dyn_cast<...>(...)` if you don't want to receive an exception when the conversion fails.

`Record` represents a TableGen record, but it would be better if we can find out its parent class, which further tells us the field's information.

For example, after getting the underlying `Record` object for `SIDef`, we can use the `isSubClassOf` function to tell if that `Record` is a baking step or ingredient, as follows:

```
Record* SIRecord = SIDef->getDef();
if (SIRecord->isSubClassOf("Step")) {
  // This Record is a baking step!
} else if (SIRecord->isSubClassOf("IngredientBase")){
  // This Record is an ingredient!
}
```

Knowing what the underlying TableGen class actually is can help us to print that record in its own way.

Handling DAG values

Now, we are going to print out the `Step` records. Recall that we used the `dag` type to represent the action and the ingredients required for a baking step. Have a look at the following code example:

```
def step_prep : Step<(heat:$action fry_oil:$oil, oil_
temp:$temp)> {
  let CustomFormat = "$action $oil until $temp";
}
```

Here, the highlighted dag is stored in the `Action` field of the `Step` TableGen class. So, we use `getValueAsDag` to retrieve that field as a `DagInit` object, as follows:

```
DagInit* DAG = StepRecord->getValueAsDag("Action");
```

`DagInit` is just another class derived from `Init`, which wasintroduced earlier. It contains some DAG-specific APIs. For example, we can iterate through all of its operands and get their associated `Init` object using the `getArg` function, as follows:

```
for(i = 0; i < DAG->arg_size; ++i) {
  Init* Arg = DAG->getArg(i);
}
```

Furthermore, we can use the `getArgNameStr` function to retrieve the token (if there is any), which is always represented in string type in the TableGen backend, associated with a specific operand, as illustrated in the following code snippet:

```
for(i = 0; i < DAG->arg_size; ++i) {
  StringRef ArgTok = DAG->getArgNameStr(i);
}
```

If `ArgTok` is empty, this means there is no token associated with that operand. To get the token associated with the operator, we can use the `getNameStr` API.

> **Note**
>
> Both `DagInit::getArgNameStr` and `DagInit::getNameStr` return the token string *without* the leading dollar sign.

This section has shown you some of the most important aspects of working with TableGen directives' in-memory C++ representation, which is the building block of writing a TableGen backend. In the next section, we will show you the final step to put everything together and run our custom TableGen backend.

Integrating the RecipePrinter TableGen backend

After finishing the `utils/TableGen/RecipePrinter.cpp` file, it's time to put everything together.

As mentioned before, a TableGen backend is always associated with the `llvm-tblgen` tool, which is also the only interface to use the backend. `llvm-tblgen` uses simple command-line options to choose a backend to use.

Here is an example of choosing one of the backends, `IntrInfoEmitter`, to generate a C/C++ header file from a `TableGen` file that carries instruction set information of X86:

```
$ llvm-tblgen -gen-instr-info /path/to/X86.td -o
GenX86InstrInfo.inc
```

Let's now see how to integrate `RecipePrinter` source file to `TableGen` backend:

1. To link the `RecipePrinter` source file into `llvm-tblgen` and add a command-line option to select it, we're going to use `utils/TableGen/TableGenBackends.h` first. This file only contains a list of TableGen backend entry functions, which are functions that take a `raw_ostream` output stream and the `RecordKeeper` object as arguments. We're also putting our `EmitRecipe` function into the list, as follows:

    ```
    ...
    void EmitX86FoldTables(RecordKeeper &RK, raw_ostream
    &OS);
    void EmitRecipe(RecordKeeper &RK, raw_ostream &OS);
    void EmitRegisterBank(RecordKeeper &RK, raw_ostream &OS);
    ...
    ```

2. Next, inside `llvm/utils/TableGen/TableGen.cpp`, we're first adding a new `ActionType` enum element and the selected command-line option, as follows:

    ```
    enum Action Type {
      ...
      GenRecipe,
    ```

```
...
}
...
cl::opt<ActionType> Action(
    cl::desc("Action to perform:"),
    cl::values(
        ...
        clEnumValN(GenRecipe, "gen-recipe",
                   "Print delicious recipes"),
        ...
    ));
```

3. After that, go to the LLVMTableGenMain function and insert the function call to EmitRecipe, as follows:

```
bool LLVMTableGenMain(raw_ostream &OS, RecordKeeper
&Records) {
  switch (Action) {
  ...
  case GenRecipe:
    EmitRecipe(Records, OS);
    break;
  }
}
```

4. Finally, don't forget to update utils/TableGen/CMakeLists.txt, as follows:

```
add_tablegen(llvm-tblgen LLVM
    ...
    RecipePrinter.cpp
    ...)
```

5. That's all there is to it! You can now run the following command:

```
$ llvm-tblgen -gen-recipe DonutRecipe.td
```

(You can optionally redirect the output to a file using the -o option.)

The preceding command will print out a (mostly) normal donut recipe, just like this:

```
=======Ingredients=======
1. oil 500 ml
2. flour 300 g
3. milk 1.25 cup
4. whole egg 1
5. yeast 1.50 tsp
6. butter 3.50 tbsp
7. sugar 2.0 tbsp
8. salt 0.50 tsp
9. vanilla extract 1.0 tsp


=======Instructions=======
1. use deep fryer to heat oil until 160 C
2. use mixer to mix flour, milk, whole egg, yeast,
butter, sugar, salt, and vanilla extract. stir in low
speed.
3. use mixer to mix outcome from (step 2). stir in medium
speed.
4. use bowl to ferment outcome from (step 3).
5. use rolling pin to flatten outcome from (step 4).
6. use cutter to cut outcome from (step 5).
7. use deep fryer to fry outcome from (step 1) and
outcome from (step 6).
```

In this section, we have learned how to build a custom TableGen backend to transform a recipe written in TableGen into normal plaintext format. Things we learned here include how `llvm-tblgen`, the driver of translating TableGen code, works; how to use the TableGen backend's C++ APIs to operate TableGen directive's in-memory representation; and how to integrate our custom backend into `llvm-tblgen` in order to run it. Combining the skills you learned in this chapter and in the previous one, you can create a complete and standalone toolchain that implements your custom logic, using TableGen as a solution.

Summary

In this chapter, we introduced TableGen, a powerful DSL for expressing structural data. We have shown you its universality in solving a variety of tasks, albeit it originally being created for compiler development. Through the lens of writing a donut recipe in TableGen, we have learned its core syntax. The following section on developing a custom TableGen backend taught you how to use C++ APIs to interact with in-memory TableGen directives parsed from the source input, giving you the power to create a complete and standalone TableGen toolchain to implement your own custom logic. Learning how to master TableGen can not only help your development in LLVM-related projects but also gives you more options to solve structural data problems in arbitrary projects.

This section marks the end of the first part—an introduction to all kinds of useful supporting components in the LLVM project. Starting from the next chapter, we will move into the core compilation pipeline of LLVM. The first important topic we will cover is Clang, LLVM's official frontend for C-family programming languages.

Further reading

- This LLVM page provides a good reference on the TableGen syntax:
  `https://llvm.org/docs/TableGen/ProgRef.html`

- This LLVM page provides a good reference on developing a TableGen backend:
  `https://llvm.org/docs/TableGen/BackGuide.html`

Section 2: Frontend Development

In this section, you will learn about topics related to the frontend, including Clang and its tooling (for example, semantic reasoning) infrastructures. We will exclusively focus on plugin developments and how to write your custom toolchains. This section includes the following chapters:

- *Chapter 5, Exploring Clang's Architecture*
- *Chapter 6, Extending the Preprocessor*
- *Chapter 7, Handling AST*
- *Chapter 8, Working with Compiler Flags and Toolchains*

5
Exploring Clang's Architecture

Clang is LLVM's official frontend for **C-family** programming languages, including C, C++, and Objective-C. It processes the input source code (parsing, type checking, and semantic reasoning, to name a few) and generates equivalent LLVM IR code, which is then taken over by other LLVM subsystems to perform optimization and native code generation. Many *C-like* dialects or language extensions also find Clang easy to host their implementations. For example, Clang provides official support for OpenCL, OpenMP, and CUDA C/C++. In addition to normal frontend jobs, Clang has been evolving to partition its functionalities into libraries and modules so that developers can use them to create all kinds of tools related to **source code processing**; for example, code refactoring, code formatting, and syntax highlighting. Learning Clang development can not only bring you more engagement into the LLVM project but also open up a wide range of possibilities for creating powerful applications and tools.

Unlike LLVM, which arranges most of its tasks into a single pipeline (that is, **PassManager**) and runs them sequentially, there is more diversity in how Clang organizes its subcomponents. In this chapter, we will show you a clear picture of how Clang's important subsystems are organized, what their roles are, and which part of the code base you should be looking for.

> **Terminology**
>
> From this chapter through to the rest of this book, we will be using Clang (which starts with an uppercase C and a Minion Pro font face) to refer to the *project* and its *techniques* as a whole. Whenever we use `clang` (all in lowercase with a Courier font face), we are referring to the *executable program*.

In this chapter, we will cover the following main topics:

- Learning Clang's subsystems and their roles
- Exploring Clang's tooling features and extension options

By the end of this chapter, you will have a roadmap of this system so that you can kickstart your own projects and have some gravity for later chapters related to Clang development.

Technical requirements

In *Chapter 1*, *Saving Resources When Building LLVM*, we showed you how to build LLVM. Those instructions, however, did not build Clang. To include Clang in the build list, please edit the value that's been assigned to the `LLVM_ENABLE_PROJECTS` CMake variable, like so:

```
$ cmake -G Ninja -DLLVM_ENABLE_PROJECTS="clang;clang-tools-extra" …
```

The value of that variable should be a semi-colon-separated list, where each item is one of LLVM's subprojects. In this case, we're including Clang and `clang-tools-extra`, which contains a bunch of useful tools based on Clang's techniques. For example, the `clang-format` tool is used by countless open source projects, especially large-scale ones, to impose a unified coding style in their code base.

> **Adding Clang to an existing build**
>
> If you already have an LLVM build where Clang was not enabled, you can edit the `LLVM_ENABLE_PROJECTS` CMake argument's value in `CMakeCache.txt` without invoking the original CMake command again. CMake should reconfigure itself once you've edited the file and run Ninja (or a build system of your choice) again.

You can build `clang`, Clang's driver, and the main program using the following command:

```
$ ninja clang
```

You can run all the Clang tests using the following command:

```
$ ninja check-clang
```

Now, you should have the `clang` executable in the `/<your build directory>/bin` folder.

Learning Clang's subsystems and their roles

In this section, we will give you an overview of Clang's structures and organizations. We will briefly introduce some of the important components or subsystems, before using dedicated sections or chapters to expand them further in later parts of this book. We hope this will give you some idea of Clang's internals and how they will benefit your development.

First, let's look at the big picture. The following diagram shows the high-level structure of Clang:

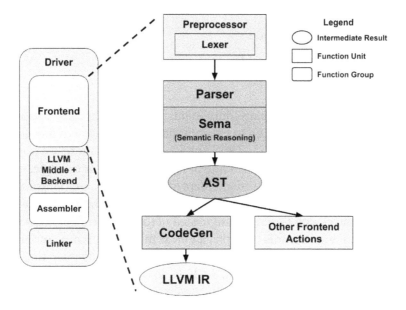

Figure 5.1 – High-level structure of Clang

As explained in the legend, rectangles with rounded corners represent subsystems that might consist of multiple components with similar functionalities. For example, the **frontend** can be further dissected into components such as the preprocessor, parser, and code generation logic, to name a few. In addition, there are intermediate results, depicted as ovals in the preceding diagram. We are especially interested in two of them – **Clang AST** and **LLVM IR**. The former will be discussed in depth in *Chapter 7, Handling AST*, while the latter is the main character of *Part 3, Middle-End Development*, which will talk about optimizations and analyses you can apply to LLVM IR.

Let's start by looking at an overview of the driver. The following subsections will give you a brief introduction to each of these driver components.

Driver

A common misunderstanding is that `clang`, the executable, is the compiler frontend. While `clang` does use Clang's frontend components, the executable itself is actually a kind of program called a **compiler driver**, or **driver** for short.

Compiling source code is a complex process. First, it consists of multiple phases, including the following:

- **Frontend**: Parsing and semantic checking
- **Middle-end**: Program analysis and optimization
- **Backend**: Native code generation
- **Assembling**: Running the assembler
- **Linking**: Running the linker

Among these phases and their enclosing components, there are countless options/arguments and flags, such as the option to tell compilers where to search for include files (that is, the `-I` command-line option in GCC and Clang). Furthermore, we hope that the compiler can figure out the values for some of these options. For example, it would be great if the compiler could include some folders of C/C++ standard libraries (for example, `/include` and `/usr/include` in Linux systems) in the header file search paths *by default*, so that we don't need to assign each of those folders manually in the command line. Continuing with this example, it's clear that we want our compilers to be **portable** across different operating systems and platforms, but many operating systems use a different C/C++ standard library path. So, how do compilers pick the correct one accordingly?

In this situation, a driver is designed to come to the rescue. It's a piece of software that acts as a housekeeper for core compiler components, serving them essential information (for example, a OS-specific system include path, as we mentioned earlier) and arranging their executions so that users only need to supply important command-line arguments. A good way to observe the hard work of a driver is to use the -### command-line flag on a normal clang invocation. For example, you could try to compile a simple hello world program with that flag:

```
$ clang++ -### -std=c++11 -Wall ./hello_world.cpp -o hello_
world
```

The following is part of the output after running the preceding command on a macOS computer:

```
"/path/to/clang" "-cc1" "-triple" "x86_64-apple-macosx11.0.0"
"-Wdeprecated-objc-isa-usage" "-Werror=deprecated-objc-
isa-usage" "-Werror=implicit-function-declaration" "-emit-
obj" "-mrelax-all" "-disable-free" "-disable-llvm-verifier"
… "-fno-strict-return" "-masm-verbose" "-munwind-tables"
"-target-sdk-version=11.0" … "-resource-dir" "/Library/
Developer/CommandLineTools/usr/lib/clang/12.0.0" "-isysroot"
"/Library/Developer/CommandLineTools/SDKs/MacOSX.sdk" "-I/
usr/local/include" "-stdlib=libc++" … "-Wall" "-Wno-reorder-
init-list" "-Wno-implicit-int-float-conversion" "-Wno-c99-
designator" … "-std=c++11" "-fdeprecated-macro" "-fdebug-
compilation-dir" "/Users/Rem" "-ferror-limit" "19"
"-fmessage-length" "87" "-stack-protector" "1" "-fstack-
check" "-mdarwin-stkchk-strong-link" … "-fexceptions" …
"-fdiagnostics-show-option" "-fcolor-diagnostics" "-o" "/path/
to/temp/hello_world-dEadBeEf.o" "-x" "c++" "hello_world.cpp"…
```

These are essentially the flags being passed to the *real* Clang frontend after the driver's *translation*. While you don't need to understand all these flags, it's true that even for a simple program, the compilation flow consists of an enormous amount of compiler options and many subcomponents.

The source code for the driver can be found under clang/lib/Driver. In *Chapter 8, Working with Compiler Flags and Toolchains*, we will look at this in more detail.

Frontend

A typical compiler textbook might tell you that a compiler frontend consists of a **lexer** and a **parser**, which generates an **abstract syntax tree** (**AST**). Clang's frontend also uses this skeleton, but with some major differences. First, the lexer is usually coupled with the **preprocessor**, and the semantic analysis that's performed on the source code is detached into a separate subsystem, called the **Sema**. This builds an AST and does all kinds of semantic checking.

Lexer and preprocessor

Due to the complexity of programming language standards and the scale of real-world source code, preprocessing becomes non-trivial. For example, resolving included files becomes tricky when you have 10+ layers of a header file hierarchy, which is common in large-scale projects. Advanced directives such as #pragma can be challenged in cases where OpenMP uses #pragma to parallelize for loops. Solving these challenges requires close cooperation between the preprocessor and the lexer, which provides primitives for all the preprocessing actions. Their source code can be found under clang/lib/Lex. In *Chapter 6, Extending the Preprocessor*, you will become familiar with preprocessor and lexer development, and learn how to implement custom logic with a powerful extension system.

Parser and Sema

Clang's parser consumes token streams from the preprocessor and lexer and tries to realize their semantic structures. Here, the Sema sub-system performs more semantic checking and analysis from the parser's result before generating the AST. Historically, there was another layer of abstraction where you could create your own *parser action* callbacks to specify what actions you wanted to perform when certain language directives (for example, identifiers such as variable names) were parsed.

Back then, Sema was one of these parser actions. However, later on, people found that this additional layer of abstraction was not necessary, so the parser only interacts with Sema nowadays. Nevertheless, Sema still retains this kind of callback-style design. For example, the clang::Sema::ActOnForStmt(...) function (defined in clang/lib/Sema/SemaStmt.cpp) will be invoked when a for loop structure is parsed. It will then do all kinds of checking to make sure the syntax is correct and generate the AST node for the for loop; that is, a ForStmt object.

AST

The AST is the most important primitive when it comes to extending Clang with your custom logic. All the common Clang extensions/plugins that we will introduce operate on an AST. To get a taste of AST, you can use the following command to print out the an AST from the source code:

```
$ clang -Xclang -ast-dump -fsyntax-only foo.c
```

For example, on my computer, I have used the following simple code, which only contains one function:

```
int foo(int c) { return c + 1; }
```

This will yield the following output:

```
TranslationUnitDecl 0x560f3929f5a8 <<invalid sloc>> <invalid
sloc>
|...
`-FunctionDecl 0x560f392e1350 <./test.c:2:1, col:30> col:5 foo
'int (int)'
  |-ParmVarDecl 0x560f392e1280 <col:9, col:13> col:13 used c
'int'
  `-CompoundStmt 0x560f392e14c8 <col:16, col:30>
    `-ReturnStmt 0x560f392e14b8 <col:17, col:28>
      `-BinaryOperator 0x560f392e1498 <col:24, col:28> 'int'
'+'
        |-ImplicitCastExpr 0x560f392e1480 <col:24> 'int'
<LValueToRValue>
        | `-DeclRefExpr 0x560f392e1440 <col:24> 'int' lvalue
ParmVar 0x560f392e1280 'c' 'int'
        `-IntegerLiteral 0x560f392e1460 <col:28> 'int' 1
```

This command is pretty useful because it tells you the C++ AST class that represents certain language directives, which is crucial for writing AST callbacks – the core of many Clang plugins. For example, from the previous lines, we can know that a variable reference site (c in the c + 1 expression) is represented by the `DeclRefExpr` class.

Similar to how the parser was organized, you can register different kinds of `ASTConsumer` instances to visit or manipulate the AST. **CodeGen**, which we will introduce shortly, is one of them. In *Chapter 7, Handling AST*, we will show you how to implement custom AST processing logic using plugins.

CodeGen

Though there are no prescriptions for how you should process the AST (for example, if you use the `-ast-dump` command-line option shown previously, the frontend will print the textual AST representation), the most common task that's performed by the CodeGen subsystem is emitting the LLVM IR code, which will later be compiled into native assembly or object code by LLVM.

LLVM, assemblers, and linkers

Once the LLVM IR code has been emitted by the CodeGen subsystem, it will be processed by the LLVM compilation pipeline to generate native code, either assembly code or object code. LLVM provides a framework called the **MC layer**, in which architectures can choose to implement assemblers that have been directly integrated into LLVM's pipeline. Major architectures such as x86 and ARM use this approach. If you don't do this, any textual assembly code that's emitted at the end of LLVM's pipeline needs to be processed by external assembler programs invoked by the driver.

Despite the fact that LLVM already has its own linker, known as the **LLD** project, an *integrated* linker is still not a mature option yet. Therefore, external linker programs are always invoked by the driver to link the object files and generate the final binary artifacts.

External versus integrated

Using external assemblers or linkers means invoking a *separate process* to run the program. For example, to run an external assembler, the frontend needs to put assembly code into a temporary file before launching the assembler with that file path as one of its command-line arguments. On the other hand, using integrated assemblers/linkers means the functionalities of assembling or linking are packaged into *libraries* rather than an executable. So, at the end of the compilation pipeline, LLVM will call APIs to process the assembly code's *in-memory* instances to emit object code. The advantage of this integrated approach is, of course, saving many indirections (writing into temporary files and reading them back right away). It also makes the code more concise to some extent.

With that, you have been given an overview of a normal compilation flow, from the source code all the way to the native code. In the next section, we will go beyond the `clang` executable and provide an overview of the tooling and extension options provided by Clang. This not only augments the functionalities of `clang`, but also provides a way to use Clang's amazing techniques in out-of-tree projects.

Exploring Clang's tooling features and extension options

The Clang project contains not just the `clang` executable. It also provides interfaces for developers to extend its tools, as well as to export its functionalities as libraries. In this section, we will give you an overview of all these options. Some of them will be covered in later chapters.

There are currently three kinds of tooling and extension options available in Clang: **Clang plugins**, **libTooling**, and **Clang Tools**. To explain their differences and provide more background knowledge when we talk about Clang extensions, we need to start from an important data type first: the `clang::FrontendAction` class.

The FrontendAction class

In the *Learning Clang's subsystems and their roles* section, we went through a variety of Clang's frontend components, such as the preprocessor and Sema, to name a few. Many of these important components are encapsulated by a single data type, called `FrontendAction`. A `FrontendAction` instance can be treated as a single task running inside the frontend. It provides a unified interface for the task to consume and interact with various resources, such as input source files and ASTs, which is similar to the role of an **LLVM Pass** from this perspective (an LLVM Pass provides a unified interface to process LLVM IR). However, there are some significant differences with an LLVM Pass:

- Not all of the frontend components are encapsulated into a `FrontendAction`, such as the parser and Sema. They are standalone components that generate materials (for example, the AST) for other FrontendActions to run.

- Except for a few scenarios (the Clang plugin is one of them), a Clang compilation instance rarely runs multiple FrontendActions. Normally, only one `FrontendAction` will be executed.

Generally speaking, a `FrontendAction` describes the task to be done at one or two important places in the frontend. This explains why it's so important for tooling or extension development – we're basically building our logic into a `FrontendAction` (one of FrontendAction's derived classes, to be more precise) instance to control and customize the behavior of a normal Clang compilation.

To give you a feel for the `FrontendAction` module, here are some of its important APIs:

- `FrontendAction::BeginSourceFileAction(…)` / `EndSourceFileAction(…)`: These are callbacks that derived classes can override to perform actions right before processing a source file and once it has been processed, respectively.

- `FrontendAction::ExecuteAction(…)`: This callback describes the main actions to do for this `FrontendAction`. Note that while no one stops you from overriding this method directly, many of FrontendAction's derived classes already provide simpler interfaces to describe some common tasks. For example, if you want to process an AST, you should inherit from `ASTFrontendAction` instead and leverage its infrastructures.

- `FrontendAction::CreateASTConsumer(…)`: This is a factory function that's used to create an `ASTConsumer` instance, which is a group of callbacks that will be invoked by the frontend when it's traversing different parts of the AST (for example, a callback to be called when the frontend encounters a group of declarations). Note that while the majority of FrontendActions work after the AST has been generated, the AST might not be generated at all. This may happen if the user only wants to run the preprocessor, for example (such as to dump the preprocessed content using Clang's `-E` command-line option). Thus, you don't always need to implement this function in your custom `FrontendAction`.

Again, normally, you won't derive your class directly from `FrontendAction`, but understanding FrontendAction's internal role in Clang and its interfaces can give you more material to work with when it comes to tooling or plugin development.

Clang plugins

A Clang plugin allows you to dynamically register a new `FrontendAction` (more specifically, an `ASTFrontendAction`) that can process the AST either before or after, or even replace, the main action of `clang`. A real-world example can be found in the **Chromium** project, in which they use Clang plugins to impose some Chromium-specific rules and make sure their code base is free from any non-ideal syntax. For example, one of the tasks is checking if the `virtual` keyword has been placed on methods that should be virtual.

A plugin can be easily loaded into a normal `clang` by using simple command-line options:

```
$ clang -fplugin=/path/to/MyPlugin.so … foo.cpp
```

This is really useful if you want to customize the compilation but have no control over the `clang` executable (that is, you can't use a modified version of `clang`). In addition, using the Clang plugin allows you to integrate with the build system more tightly; for example, if you want to rerun your logic once the source files or even arbitrary build dependencies have been modified. Since the Clang plugin is still using `clang` as the driver and modern build systems are pretty good at resolving normal compilation command dependencies, this can be done by making a few compile flag tweaks.

However, the biggest downside of using the Clang plugin is its **API issue**. In theory, you can load and run your plugin in any `clang` executable, but only if the C++ APIs (and the ABI) are used by your plugin and the `clang` executable matches it. Unfortunately, for now, Clang (and also the whole LLVM project) has no intention to make any of its C++ APIs stable. In other words, to take the safest path, you need to make sure both your plugin and `clang` are using the *exact same* (major) version of LLVM. This issue makes the Clang plugin pretty hard to be released standalone.

We will look at this in more detail in *Chapter 7, Handling AST*.

LibTooling and Clang Tools

LibTooling is a library that provides features for building *standalone tools* on top of Clang's techniques. You can use it like a normal library in your project, without having any dependencies on the `clang` executable. Also, the APIs are designed to be more high-level so that you don't need to deal with many of Clang's internal details, making it more friendly to non-Clang developers.

Language server is one of the most famous use cases of libTooling. A Language server is launched as a daemon process and accepts requests from editors or IDEs. These requests can be as simple as syntax checking a code snippet or complicated tasks such as code completions. While a Language server does not need to compile the incoming source code into native code as normal compilers do, it needs a way to parse and analyze that code, which is non-trivial to build from scratch. libTooling avoids the need to *recreate the wheels* in this case by taking Clang's techniques off-the-shelf and providing an easier interface for Language server developers.

To give you a more concrete idea of how libTooling differs from the Clang plugin, here is a (simplified) code snippet for executing a custom `ASTFrontendAction` called `MyCustomAction`:

```
int main(int argc, char** argv) {
    CommonOptionsParser OptionsParser(argc, argv,…);
    ClangTool Tool(OptionsParser.getCompilations(), {"foo.cpp"});
```

```
return Tool.run(newFrontendActionFactory<MyCustomAction>().
    get());
}
```

As shown in the previous code, you can't just embed this code into *any* code base. libTooling also provides lots of nice utilities, such as `CommonOptionsParser`, which parses textual command-line options and transforms them into Clang options for you.

> **libTooling's API Stability**
>
> Unfortunately, libTooling doesn't provide stable C++ APIs either. Nevertheless, this isn't a problem since you have full control over what LLVM version you're using.

Last but not least, **Clang Tools** is a collection of utility programs build on top of libTooling. You can think of it as the command-line tool version of libTooling in that it provides some common functionalities. For example, you can use `clang-refactor` to refactor the code. This includes renaming a variable, as shown in the following code:

```cpp
// In foo.cpp...
struct Location {
  float Lat, Lng;
};
float foo(Location *loc) {
  auto Lat = loc->Lat + 1.0;
  return Lat;
}
```

If we want to rename the `Lat` member variable in the `Location` struct `Latitude`, we can use the following command:

```
$ clang-refactor --selection="foo.cpp:1:1-10:2" \
                 --old-qualified-name="Location::Lat" \
                 --new-qualified-name="Location::Latitude" \
                 foo.cpp
```

> **Building clang-refactor**
>
> Be sure to follow the instructions at the beginning of this chapter to include `clang-tools-extra` in the list for the `LLVM_ENABLE_PROJECTS` CMake variable. By doing this, you'll be able to build `clang-refactor` using the `ninja clang-refactor` command.

You will get the following output:

```
// In foo.cpp…
struct Location {
  float Latitude, Lng;
};
float foo(Location *loc) {
  auto Lat = loc->Latitude + 1.0;
  return Lat;
}
```

This is done by the refactoring framework built inside libTooling; `clang-refactor` merely provides a command-line interface for it.

Summary

In this chapter, we looked at how Clang is organized and the functionalities of some of its important subsystems and components. Then, we learned about the differences between Clang's major extension and tooling options – the Clang plugin, libTooling, and Clang Tools – including what each of them looks like and what their pros and cons are. The Clang plugin provides an easy way to insert custom logic into Clang's compilation pipeline via dynamically loaded plugins but suffers from API stability issues; libTooling has a different focus than the Clang plugin in that it aims to provide a toolbox for developers to create a standalone tool; and Clang Tools provides various applications.

In the next chapter, we will talk about preprocessor development. We will learn how the preprocessor and the lexer work in Clang, and show you how to write plugins for the sake of customizing preprocessing logic.

Further reading

- Here is a list of checks that are done by Chromium's Clang plugin: `https://chromium.googlesource.com/chromium/src/tools/clang/+/refs/heads/master/plugins/FindBadConstructsAction.h`.

- You can learn more about choosing the right Clang extension interface here: `https://clang.llvm.org/docs/Tooling.html`.

- LLVM also has its own libTooling-based Language server, called `clangd`: `http://clangd.llvm.org`.

6
Extending the Preprocessor

In the previous chapter, we went through the structure of Clang—the official frontend of **Low-Level Virtual Machine (LLVM)** for C-family languages—and some of its most important components. We also introduced a variety of Clang's tooling and extension options. In this chapter, we're diving into the first phase in Clang's frontend pipeline: the preprocessor.

For C-family programming languages, **preprocessing** is an early compilation phase that replaces any directive starting with a hash (#) character—#include and #define, to name but a few—with some other textual contents (or non-textual *tokens*, in some rare cases). For example, the preprocessor will basically *copy and paste* contents of header files designated by the #include directive into the current compilation unit before parsing it. This technique has the advantage of extracting common code and reusing it.

In this chapter, we will briefly explain how Clang's **preprocessor/Lexer** framework works, along with some crucial **application programming interfaces** (**APIs**) that can help your development in this section. In addition, Clang also provides some ways for developers to inject their custom logic into the preprocessing flow via plugins. For example, it allows you to create custom #pragma syntax—such as that used by OpenMP (#pragma omp loop, for example) —in an easier way. Learning these techniques yields you more options when solving problems of different abstraction levels. Here is the list of sections in this chapter:

- Working with SourceLocation and SourceManager
- Learning preprocessor and lexer essentials
- Developing custom preprocessor plugins and callbacks

Technical requirements

This chapter expects you to have a build of the Clang executable. You can obtain this by running the following command:

```
$ ninja clang
```

Here is a useful command to print textual content right after preprocessing:

The -E command-line option for clang is pretty useful for printing textual content right after preprocessing. As an example, foo.c has the following content:

```
#define HELLO 4
int foo(int x) {
  return x + HELLO;
}
```

Use the following command:

```
$ clang -E foo.c
```

The preceding command will give you this output:

```
...
int foo(int x) {
  return x + 4;
}
```

As you can see, HELLO was replaced by 4 in the code. You might be able to use this trick to debug when developing custom extensions in later sections.

Code used in this chapter can be found at this link: https://github.com/ PacktPublishing/LLVM-Techniques-Tips-and-Best-Practices-Clang- and-Middle-End-Libraries/tree/main/Chapter06.

Working with SourceLocation and SourceManager

When working closely with source files, one of the most fundamental questions is how a compiler frontend would be able to *locate* a piece of string in the file. On one hand, printing format messages well (compilation error and warning messages, for example) is a crucial job, in which accurate line and column numbers must be displayed. On the other hand, the frontend might need to manage multiple files at a time and access their in-memory content in an efficient way. In Clang, these questions are primarily handled by two classes: SourceLocation and SourceManager. We're going to give you a brief introduction to them and show how to use them in practice in the rest of this section.

Introducing SourceLocation

The SourceLocation class is used for representing the location of a piece of code in its file. When it comes to its implementation, using **line** and **column** numbers is probably the most intuitive way to do this. However, things might get complicated in real-world scenarios, such that internally, we can't naively use a pair of numbers as the in-memory representations for source code locations. One of the main reasons is that SourceLocation instances are *extensively* used in Clang's code base and basically live through the entire frontend compilation pipeline. Therefore, it's important to use a concise way to store its information rather than two 32-bit integers (and this might not even be sufficient since we also want to know the origin file!), which can easily bloat Clang's runtime-memory footprint.

Clang solves this problem by using the elegantly designed SourceLocation as the **pointer** (or a *handle*) to a large *data buffer* that stores all the real source code contents such that SourceLocation only uses a single unsigned integer under the hood, which also means its instances are **trivially copyable**—a property that can yield some performance benefits. Since SourceLocation is merely a pointer, it will only be meaningful and useful when put side by side with the *data buffer* we just mentioned, which is managed by the second main character in this story, SourceManager.

> **Other useful utilities**
>
> SourceRange is a pair of SourceLocation objects that represents
> the starting and ending of a source code range; FullSourceLocation
> encapsulates the normal SourceLocation class and its associated
> SourceManager class into one class so that you only need to carry
> a single FullSourceLocation instance instead of two objects
> (a SourceLocation object and a SourceManager object).

Trivially copyable

We were usually taught that unless there is a good reason, you should avoid passing an object by its value (as a function call argument, for example) in normal situations when writing C++. Since it involves lots of *copying* on the data members under the hood, you should pass by pointers or references instead. However, if carefully designed, a class type instance can be copied back and forth without lots of effort—for example, a class with no member variable or few member variables, plus a default copy constructor. If an instance is trivially copyable, you're encouraged to pass it by its value.

Introducing SourceManager

The SourceManager class manages all of the source files stored inside the memory and provides interfaces to access them. It also provides APIs to deal with source code locations, via SourceLocation instances we just introduced. For example, to get the line and column number from a SourceLocation instance, run the following code:

```
void foo(SourceManager &SM, SourceLocation SLoc) {
  auto Line = SM.getSpellingLineNumber(SLoc),
       Column = SM.getSpellingColumnNumber(SLoc);
  …
}
```

The Line and Column variables in the preceding code snippet are the line and column number of the source location pointed by SLoc, respectively.

You might wonder why we are using the term `spellingLineNumber` instead of just `LineNumber` in the preceding code snippet. It turns out that in the cases of macro expansion (or any expansion happening during preprocessing), Clang keeps track of the macro content's `SourceLocation` instance before and after the expansion. A spelling location represents the location where the source code was originally *written*, whereas an expansion location is where the macro is expanded.

You can also create a new spelling and expansion association using the following API:

```
SourceLocation NewSLoc = SM.createExpansionLoc(
  SpellingLoc,     // The original macro spelling location
  ExpansionStart,  // Start of the location where macro is
                   //expanded
  ExpansionEnd,    // End of the location where macro is
                   // expanded
  Len              // Length of the content you want to expand
);
```

The returned `NewSLoc` is now associated with both the spelling and expanded locations that can be queried using `SourceManager`.

These are the important concepts and APIs that will help you dealing with source code locations— especially when working with the preprocessor—in later chapters. The next section will give you some background on preprocessor and lexer development in Clang, which will be useful when working on the project in the later, *Developing custom preprocessor plugins and callbacks* section.

Learning preprocessor and lexer essentials

In the previous, *Working with SourceLocation and SourceManager* section, we've learned how source locations, which are an important part of the preprocessor, are represented in Clang. In this section, we will first explain the principle of Clang's preprocessor and lexer, along with their working flow. Then, we'll go into some of the important components in this flow and briefly explain their usage in the code. These will also prepare you for the project in the, *Developing custom preprocessor plugins and callbacks* section later in this chapter.

Understanding the role of the preprocessor and lexer in Clang

The roles and primary actions performed by Clang's preprocessor and lexer, represented by the `Preprocessor` and `Lexer` classes respectively, are illustrated in the following diagram:

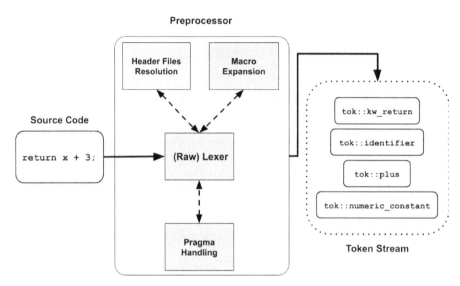

Figure 6.1 – Role of the Clang preprocessor and lexer

We believe most readers here will be familiar with the concept of a **token** in the context of the lexer—a substring from the original source code that acts as the minimum building block for semantic reasoning. In some of the traditional compilers, the lexer is responsible for chopping the input source code into a sequence of tokens or a **token stream**, as shown in the preceding diagram. This token stream will later be fed into the parser to construct the semantic structure.

Implementation-wise, Clang takes a slightly different path from traditional compilers (or those from textbooks): `Lexer`, employed by `Preprocessor`, is still the primary performer to cut source code into tokens. However, `Lexer` keeps its hands off whenever encountering a preprocessor directive (that is, anything that starts with a #) or a symbol, and relays that task to either the macro expansion, the header file resolver, or pragma handlers that are organized by the `Preprocessor`. These assisting components inject extra tokens, if needed, into the main token stream, which would eventually be returned back to the user of `Preprocessor`.

In other words, most consumers of the token stream don't directly interact with `Lexer`, but with the `Preprocessor` instances. This makes people call the `Lexer` class a *raw* lexer (as shown in the previous diagram), since `Lexer` by itself only generates a token stream that hasn't been preprocessed. To give you a more concrete idea of how to use `Preprocessor` to retrieve a token (stream), the following simple code snippet has been provided. This shows a way to get the next token from the source code currently processing it:

```
Token GetNextToken(Preprocessor &PP) {
  Token Tok;
  PP.Lex(Tok);
  return Tok;
}
```

As you might have guessed, `Token` is the class representing a single token in Clang, which we're going to introduce shortly in the next paragraph.

Understanding Token

The `Token` class is the representation of a single token, either from the source code or a *virtual* one that served a special purpose. It is also used extensively by the preprocessing/lexing framework, just like `SourceLocation` that we introduced earlier. Thus, it is designed to be very concise in memory and trivially copyable as well.

For the `Token` class, there are two things we want to highlight here, as follows:

1. **Token kind** tells you what this token is.
2. **Identifier** represents both language keywords and arbitrary frontend tokens (a function name, for example). Clang's preprocessor used a dedicated `IdentifierInfo` class to carry extra identifier information, which we're going to cover later in this section.

Token kind

The token kind tells you *what* this `Token` is. Clang's `Token` is designed to represent not just concrete, physical-language constructions such as keywords and symbols, but also virtual concepts that are inserted by the parser in order to encode as much information as possible using a single `Token`. To visualize the token stream's token kinds, you can use the following command-line option:

```
$ clang -fsyntax-only -Xclang -dump-tokens foo.cc
```

`foo.cc` has the following content:

```
namespace foo {
  class MyClass {};
}
foo::MyClass Obj;
```

This is the output of the preceding command:

```
namespace 'namespace'    [StartOfLine]  Loc=<foo.cc:1:1>
identifier 'foo'         [LeadingSpace] Loc=<foo.cc:1:11>
l_brace '{'      [LeadingSpace] Loc=<foo.cc:1:15>
class 'class'    [StartOfLine] [LeadingSpace]   Loc=<foo.
cc:2:3>
identifier 'MyClass'       [LeadingSpace] Loc=<foo.cc:2:9>
l_brace '{'      [LeadingSpace] Loc=<foo.cc:2:17>
r_brace '}'              Loc=<foo.cc:2:18>
semi ';'                 Loc=<foo.cc:2:19>
r_brace '}'      [StartOfLine]  Loc=<foo.cc:3:1>
identifier 'foo'          [StartOfLine]  Loc=<foo.cc:5:1>
coloncolon '::'          Loc=<foo.cc:5:4>
identifier 'MyClass'        Loc=<foo.cc:5:6>
identifier 'Obj'          [LeadingSpace] Loc=<foo.cc:5:14>
semi ';'                 Loc=<foo.cc:5:17>
eof ''            Loc=<foo.cc:5:18>
```

The highlighted parts are the token kinds for each token. The full list of token kinds can be found in `clang/include/clang/Basic/TokenKinds.def`. This file is a useful reference to know the mapping between any language construction (for example, the `return` keyword) and its token kind counterpart (`kw_return`).

Although we can't visualize the virtual tokens—or **annotation tokens**, as they are called in Clang's code base—we will still explain these using the same example as before. In C++, :: (the `coloncolon` token kind in the preceding directive) has several different usages. For example, it can either be for namespace resolution (more formally called *scope resolution* in C++), as shown in the code snippet earlier, or it can be (optionally) used with the `new` and `delete` operators, as illustrated in the following code snippet:

```
int* foo(int N) {
    return ::new int[N]; // Equivalent to 'new int[N]'
}
```

To make the parsing processing more efficient, the parser will first try to resolve whether the `coloncolon` token is a scope resolution or not. If it is, the token will be replaced by an `annot_cxxscope` annotation token.

Now, let's see the API to retrieve the token kind. The `Token` class provides a `getKind` function to retrieve its token kind, as illustrated in the following code snippet:

```
bool IsReturn(Token Tok) {
    return Tok.getKind() == tok::kw_return;
}
```

However, if you're only doing checks, just like in the preceding snippet, a more concise function is available, as illustrated here:

```
bool IsReturn(Token Tok) {
    return Tok.is(tok::kw_return);
}
```

Though many times, knowing the token kind of a `Token` is sufficient for processing, some language structures require more evidence to judge (for example, tokens that represent a function name, in which case the token kind, `identifier`, is not as important as the name string). Clang uses a specialized class, `IdentifierInfo`, to carry extra information such as the symbol name for any identifier in the language, which we're going to cover in the next paragraph.

Identifier

Standard C/C++ uses the word **identifier** to represent a wide variety of language concepts, ranging from symbol names (such as function or macro names) to language keywords, which are called **reserved identifiers** by the standard. Clang also follows a similar path on the implementation side: it decorates `Token` that fit into the language's standard definition of an identifier with an auxiliary `IdentifierInfo` object. This object encloses properties such as the underlying string content or whether this identifier is associated with a macro function. Here is how you would retrieve the `IdentifierInfo` instance from a `Token` type variable `Tok`:

```
IdentifierInfo *II = Tok.getIdentifierInfo();
```

The preceding `getIdentifierInfo` function returns null if `Tok` is not representing an identifier by the language standard's definition. Note that if two identifiers have the *same textual content*, they are represented by the same `IdentifierInfo` object. This comes in handy when you want to compare whether different identifier tokens have the same textual contents.

Using a dedicated `IdentifierInfo` type on top of various token kinds has the following advantages:

- For a `Token` with an `identifier` token kind, we sometimes want to know if it has been associated with a macro. You can find this out with the `IdentifierInfo::hasMacroDefinition` function.

- For a token with an `identifier` token kind, storing underlying string content in auxiliary storage (that is, the `IdentifierInfo` object) can save a `Token` object's memory footprint, which is on the hot path of the frontend. You can retrieve the underlying string content with the `IdentifierInfo::getName` function.

- For a `Token` that represents language keywords, though the framework already provides dedicated token kinds for these sorts of tokens (for example, `kw_return` for the `return` keyword), some of these tokens only become language keywords in later language standards. For example, the following snippet is legal in standards before C++11:

  ```
  void foo(int auto) {}
  ```

- You could compile it with the following command:

  ```
  $ clang++ -std=c++03 -fsyntax-only …
  ```

If you do so, it won't give you any complaint, until you change the preceding `-std=c++03` standard into `-std=c++11` or a later standard. The error message in the latter case will say that `auto`, a language keyword since C++11, can't be used there. To give the frontend have an easier time judging if a given token is a keyword in any case, the `IdentifierInfo` object attached on keyword tokens is designed to answer if an identifier is a keyword under a certain language standard (or language feature), using the `IdentifierInfo::isKeyword(...)` function, for example, whereby you pass a `LangOptions` class object (a class carrying information such as the language standard and features currently being used) as the argument to that function.

In the next sub-section, we're going to introduce the last important `Preprocessor` concept of this section: how `Preprocessor` handles *macros* in C-family languages.

Handling macros

Implementations for **macros** of C-family languages are non-trivial. In addition to challenges on source locations as we introduced earlier—how do we carry source locations of both the macro definitions and the place they're expanded—the ability to re-define and undefine a macro name complicates the whole story. Have a look at the following code snippet for an example of this:

```
#define FOO(X) (X + 1)
return FOO(3); // Equivalent to "return (3 + 1);"
#define FOO(X) (X - 100)
return FOO(3); // Now this is equivalent to "return (3 - 100);"
#undef FOO
return FOO(3); // "FOO(3)" here will not be expanded in
               //preprocessor
```

The preceding C code showed that the definition of FOO (if FOO is defined) varies on different lexical locations (different lines).

> **Local versus Module macros**
>
> C++20 has introduced a new language concept called **Module**. It resembles the modularity mechanisms in many other object-oriented languages such as Java or Python. You can also define macros in a Module, but they work slightly differently from the traditional macros, which are called **local macros** in Clang. For example, you can control the visibility of a Module macro by using keywords such as `export`. We only cover local macros in this book.

To model this concept, Clang has constructed a system to record the chain of definitions and un-definitions. Before explaining how it works, here are three of the most important components of this system:

1. `MacroDirective`: This class is the logical representation of a #define or a #undef *statement* of a given macro identifier. As shown in the preceding code example, there can be multiple #define (and #undef) statements on the same macro identifier, so eventually these `MacroDirective` objects will form a chain ordered by their lexical appearances. To be more specific, the #define and #undef directives are actually represented by subclasses of `MacroDirective`, `DefMacroDirective`, and `UndefMacroDirective`, respectively.

2. `MacroDefinition`: This class represents the *definition* of a macro identifier at the current time point. Rather than containing the full macro definition body, this instance is more like a pointer pointing to different macro bodies, which are represented by the `MacroInfo` class that will be introduced shortly, upon resolving a different `MacroDirective` class. This class can also tell you the (latest) `DefMacroDirective` class that defines this `MacroDefinition` class.

3. `MacroInfo`: This class contains the body, including tokens in the body and macro arguments (if any) of a macro definition.

Here is a diagram illustrating the relationship of these classes in regard to the sample code earlier:

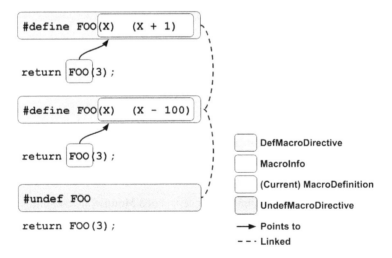

Figure 6.2 – How different C++ classes for a macro are related to the previous code example

To retrieve the `MacroInfo` class and its `MacroDefinition` class, we can use the following `Preprocessor` APIs, as follows:

```
void printMacroBody(IdentifierInfo *MacroII, Preprocessor &PP)
{
  MacroDefinition Def = PP.getMacroDefinition(MacroII);
  MacroInfo *Info = Def.getMacroInfo();

  …
}
```

The `IdentifierInfo` type argument, `MacroII`, shown in the preceding code snippet, represents the macro name. To further examine the macro body, run the following code:

```
void printMacroBody(IdentifierInfo *MacroII, Preprocessor &PP)
{

  …
  MacroInfo *Info = Def.getMacroInfo();
  for(Token Tok : Info->tokens()) {
    std::cout << Tok.getName() << "\n";
  }
}
```

From this section, you've learned the working flow of `Preprocessor`, as well as two important components: the `Token` class and the sub-system that handles macros. Learning these two gives you a better picture of how Clang's preprocessing works and prepares you for the `Preprocessor` plugin and custom callbacks development in the next section.

Developing custom preprocessor plugins and callbacks

As flexible as other parts of LLVM and Clang, Clang's preprocessing framework also provides a way to insert custom logic via plugins. More specifically, it allows developers to write plugins to handle custom **pragma** directives (that is, allowing users to write something such as #pragma my_awesome_feature). In addition, the `Preprocessor` class also provides a more general way to define custom callback functions in reaction to arbitrary **preprocessing events**— such as when a macro is expanded or a #include directive is resolved, to name but a couple of examples. In this section, we're going to use a simple project that leverages both techniques to demonstrate their usage.

The project goal and preparation

Macros in C/C++ have always been notorious for poor *design hygiene* that could easily lead to coding errors when used without care. Have a look at the following code snippet for an example of this:

```
#define PRINT(val) \
    printf("%d\n", val * 2)
void main() {
    PRINT(1 + 3);
}
```

PRINT in the preceding code snippet looks just like a normal function, thus it's easy to believe that this program will print out 8. However, PRINT is a macro function rather than a normal function, so when it's expanded, the main function is equivalent to this:

```
void main() {
    printf("%d\n", 1 + 3 * 2);
}
```

Therefore, the program actually prints 7. This ambiguity can of course be solved by wrapping every occurrence of the val macro argument in the macro body with parenthesis, as illustrated in the following code snippet:

```
#define PRINT(val) \
    printf("%d\n", (val) * 2)
```

Therefore, after macro expansion, the main function will look like this:

```
void main() {
    printf("%d\n", (1 + 3) * 2);
}
```

The project we're going to do here is to develop a custom #pragma syntax to warn developers if a certain macro argument, designated by programmers, is not properly enclosed in parentheses, for the sake of preventing the preceding *hygiene* problems from happening. Here is an example of this new syntax:

```
#pragma macro_arg_guard val
#define PRINT(val) \
    printf("%d\n", val * 94 + (val) * 87);
void main() {
```

```
    PRINT(1 + 3);
}
```

Similar to previous example, if an occurrence of the preceding `val` argument is not enclosed in parentheses, this might introduce potential bugs.

In the new `macro_arg_guard` pragma syntax, tokens following the pragma name are the macro argument names to check in the next macro function. Since `val` in the `val *` `94` expression from the preceding code snippet is not enclosed in parentheses, it will print the following warning message:

```
$ clang … foo.c
[WARNING] In foo.c:3:18: macro argument 'val' is not enclosed
by parenthesis
```

This project, albeit being a *toy example*, is actually pretty useful when the macro function becomes pretty big or complicated, in which case manually adding parentheses on *every* macro argument occurrence might be an error-prone task. A tool to catch this kind of mistake would definitely be helpful.

Before we dive into the coding part, let's set up the project folder. Here is the folder structure:

```
MacroGuard
    |___ CMakeLists.txt
    |___ MacroGuardPragma.cpp
    |___ MacroGuardValidator.h
    |___ MacroGuardValidator.cpp
```

The `MacroGuardPragama.cpp` file includes a custom `PragmaHandler` function, which we're going to cover in the next section, *Implementing a custom pragma handler*. For `MacroGuardValidator.h/.cpp`, this includes a custom `PPCallbacks` function used to check if the designated macro body and arguments conform to our rules here. We will introduce this in the later, *Implementing custom preprocessor callbacks* section.

Since we're setting up an out-of-tree project here, please refer to the *Understanding CMake integration for out-of-tree projects* section of *Chapter 2, Exploring LLVM's Build System Features*, in case you don't know how to import LLVM's own CMake directives (such as the add_llvm_library and add_llvm_executable CMake functions). And because we're also dealing with Clang here, we need to use a similar way to import Clang's build configurations, such as the include folder path shown in the following code snippet:

```
# In MacroGuard/CmakeLists.txt

…
# (after importing LLVM's CMake directives)
find_package(Clang REQUIRED CONFIG)
include_directories(${CLANG_INCLUDE_DIRS})
```

The reason we don't need to set up Clang's library path here is because normally, plugins will dynamically link against libraries' implementations provided by the loader program (in our case, the clang executable) rather than linking those libraries explicitly during build time.

Finally, we're adding the plugin's build target, as follows:

```
set(_SOURCE_FILES
    MacroGuardPragma.cpp
    MacroGuardValidator.cpp
    )
add_llvm_library(MacroGuardPlugin MODULE
                 ${_SOURCE_FILES}
                 PLUGIN_TOOL clang)
```

The PLUGIN_TOOL argument

The PLUGIN_TOOL argument for the add_llvm_library CMake function seen in the preceding code snippet is actually designed exclusively for Windows platforms, since **dynamic link library** (**DLL**) files—the dynamic shared object file format in Windows—has an…interesting rule that requires a loader executable's name to be shown in the DLL file header. PLUGIN_TOOL is also used for specifying this plugin loader executable's name.

After setting up the CMake script and building the plugin, you can use the following command to run the plugin:

```
$ clang … -fplugin=/path/to/MacroGuardPlugin.so foo.c
```

Of course, we haven't currently written any code, so nothing is printed out. In the next section, we will first develop a custom PragmaHandler instance to implement our new #pragma macro_arg_guard syntax.

Implementing a custom pragma handler

The first step of implementing the aforementioned features is to create a custom #pragma handler. To do so, we first create a MacroGuardHandler class that derives from the PragmaHandler class inside the MacroGuardPragma.cpp file, as follows:

```
struct MacroGuardHandler : public PragmaHandler {
  MacroGuardHandler() : PragmaHandler("macro_arg_guard"){}
  void HandlePragma(Preprocessor &PP, PragmaIntroducer
                    Introducer, Token &PragmaTok) override;
};
```

The HandlePragma callback function will be invoked whenever the Preprocessor encounters a non-standard pragma directive. We're going to do two things in this function, as follows:

1. Retrieve any supplement tokens—treated as the **pragma arguments**—that follows after the pragma name token (macro_arg_guard).

2. Register a PPCallbacks instance that scans the body of the next macro function definition to see if specific macro arguments are properly enclosed by parentheses in there. We will outline the details of this task next.

For the first task, we are leveraging Preprocessor to help us parse the pragma arguments, which are macro argument names to be enclosed. When HandlePragma is called, the Preprocessor is stopped at the place right after the pragma name token, as illustrated in the following code snippet:

```
#pragma macro_arg_guard val
                        ^--Stop at here
```

So, all we need to do is keep lexing and storing those tokens until hitting the end of this line:

```
void MacroGuardHandler::HandlePragma(Preprocessor &PP,…) {
  Token Tok;
  PP.Lex(Tok);
  while (Tok.isNot(tok::eod)) {
    ArgsToEnclosed.push_back(Tok.getIdentifierInfo());
    PP.Lex(Tok);
  }
}
```

The eod token kind in the preceding code snippet means **end of directive**. It is exclusively used to mark the end of a preprocessor directive.

For the ArgsToEscped variable, the following global array stores the designated macro argument's IdentifierInfo objects:

```
SmallVector<const IdentifierInfo*, 2> ArgsToEnclosed;
struct MacroGuardHandler: public PragmaHandler {
  …
};
```

The reason we're declaring ArgsToEnclosed in a global scope is that we're using it to *communicate* with our PPCallbacks instance later, which will use that array content to perform the validations.

Though the implementation details of our PPCallbacks instance, the MacroGuardValidator class, will not be covered until the next section, it needs to be registered with the Preprocessor when the HandlePragma function is called for the first time, as follows:

```
struct MacroGuardHandler : public PragmaHandler {
  bool IsValidatorRegistered;
  MacroGuardHandler() : PragmaHandler("macro_arg_guard"),
                        IsValidatorRegistered(false) {}
  …
};
void MacroGuardHandler::HandlePragma(Preprocessor &PP,…) {
  …
```

```
  if (!IsValidatorRegistered) {
    auto Validator = std::make_unique<MacroGuardValidator>(…);
    PP.addCallbackPPCallbacks(std::move(Validator));
    IsValidatorRegistered = true;
  }
}
```

We also use a flag to make sure it is only registered once. After this, whenever a preprocessing event happens, our `MacroGuardValidator` class will be invoked to handle it. In our case, we are only interested in the `macro definition` event, which signals to `MacroGuardValidator` to validate the macro body that it just defined.

Before wrapping up on `PragmaHandler`, we need some extra code to transform the handler into a plugin, as follows:

```
struct MacroGuardHandler : public PragmaHandler {
  …
};
static PragmaHandlerRegistry::Add<MacroGuardHandler>
  X("macro_arg_guard", "Verify if designated macro args are
    enclosed");
```

After declaring this variable, when this plugin is loaded into `clang`, a `MacroGuardHandler` instance is inserted into a global `PragmaHandler` registry, which will be queried by the `Preprocessor` whenever it encounters a non-standard `#pragma` directive. Now, Clang is able to recognize our custom `macro_arg_guard` pragma when the plugin is loaded.

Implementing custom preprocessor callbacks

`Preprocessor` provides a set of callbacks, the `PPCallbacks` class, which will be triggered when certain preprocessor events (such as a macro being expanded) happen. The previous, *Implementing a custom pragma handler* section, showed you how to register your own `PPCallbacks` implementations, the `MacroGuardValidator`, with `Preprocessor`. Here, we're going to show you how `MacroGuardValidator` validates the macro argument-escaping rule in macro functions.

First, in `MacroGuardValidator.h`/`.cpp`, we put the following skeleton:

```cpp
// In MacroGuardValidator.h
extern SmallVector<const IdentifierInfo*, 2> ArgsToEnclosed;

class MacroGuardValidator : public PPCallbacks {
  SourceManager &SM;
public:
  explicit MacroGuardValidator(SourceManager &SM) : SM(SM) {}
  void MacroDefined(const Token &MacroNameToke,
                    const MacroDirective *MD) override;
};

// In MacroGuardValidator.cpp
void MacroGuardValidator::MacroDefined(const Token
&MacroNameTok, const MacroDirective *MD) {
}
```

Among all the callback functions in `PPCallbacks`, we're only interested in `MacroDefined`, which will be invoked when a macro definition is processed, represented by the `MacroDirective` type function argument (`MD`). The `SourceManager` type member variable (`SM`) is used for printing `SourceLocation` when we need to show some warning messages.

Focusing on `MacroGuardValidator::MacroDefined`, the logic here is pretty simple: for each identifier in the `ArgsToEnclosed` array, we're scanning macro body tokens to check if its occurrences have parentheses as its predecessor and successor tokens. First, let's put in the loop's skeleton, as follows:

```cpp
void MacroGuardValidator::MacroDefined(const Token
&MacroNameTok, const MacroDirective *MD) {
  const MacroInfo *MI = MD->getMacroInfo();
  // For each argument to be checked...
  for (const IdentifierInfo *ArgII : ArgsToEnclosed) {
    // Scanning the macro body
    for (auto TokIdx = 0U, TokSize = MI->getNumTokens();
         TokIdx < TokSize; ++TokIdx) {
      ...
    }
```

```
            }
    }
```

If a macro body token's `IdentifierInfo` argument matches `ArgII`, this means there is a macro argument occurrence, and we check that token's previous and next tokens, as follows:

```
for (const IdentifierInfo *ArgII : ArgsToEnclosed) {
    for (auto TokIdx = 0U, TokSize = MI->getNumTokens();
         TokIdx < TokSize; ++TokIdx) {
      Token CurTok = *(MI->tokens_begin() + TokIdx);
      if (CurTok.getIdentifierInfo() == ArgII) {
        if (TokIdx > 0 && TokIdx < TokSize - 1) {
          auto PrevTok = *(MI->tokens_begin() + TokIdx - 1),
               NextTok = *(MI->tokens_begin() + TokIdx + 1);
          if (PrevTok.is(tok::l_paren) && NextTok.is
            (tok::r_paren))
            continue;
        }
        ...
      }
    }
}
```

Uniqueness of `IdentifierInfo` instances

Recall that same identifier strings are always represented by the same `IdentifierInfo` object. That's the reason we can simply use pointer comparison here.

The `MacroInfo::tokens_begin` function returns an iterator pointing to the beginning of an array carrying all the macro body tokens.

Finally, we print a warning message if the macro argument token is not enclosed by parentheses, as follows:

```
for (const IdentifierInfo *ArgII : ArgsToEnclosed) {
    for (auto TokIdx = 0U, TokSize = MI->getNumTokens();
         TokIdx < TokSize; ++TokIdx) {
      ...
```

```
    if (CurTok.getIdentifierInfo() == ArgII) {
      if (TokIdx > 0 && TokIdx < TokSize - 1) {
        ...
        if (PrevTok.is(tok::l_paren) && NextTok.is
          (tok::r_paren))
          continue;
      }
      SourceLocation TokLoc = CurTok.getLocation();
      errs() << "[WARNING] In " << TokLoc.printToString(SM)
            << ": ";
      errs() << "macro argument '" << ArgII->getName()
            << "' is not enclosed by parenthesis\n";
    }
  }
}
```

And that's all for this section. You're now able to develop a PragmaHandler plugin that can be dynamically loaded into Clang to handle custom #pragma directives. You've also learned how to implement PPCallbacks to insert custom logic whenever a preprocessor event happens.

Summary

The preprocessor and lexer mark the beginning of a frontend. The former replaces preprocessor directives with other textual contents, while the latter cuts source code into more meaningful tokens. In this chapter, we've learned how these two components cooperate with each other to provide a single view of token streams to work on in later stages. In addition, we've also learned about various important APIs—such as the Preprocessor class, the Token class, and how macros are represented in Clang—that can be used for the development of this part, especially for creating handler plugins to support custom #pragma directives, as well as creating custom preprocessor callbacks for deeper integration with preprocessing events.

Following the order of Clang's compilation stages, the next chapter will show you how to work with an **abstract syntax tree** (**AST**) and how to develop an AST plugin to insert custom logic into it.

Exercises

Here are some simple questions and exercises that you might want to play around with by yourself:

1. Though most of the time `Tokens` are harvested from provided source code, in some cases, `Tokens` might be generated dynamically inside the `Preprocessor`. For example, the `__LINE__` built-in macro is expanded to the current line number, and the `__DATE__` macro is expanded to the current calendar date. How does Clang put that generated textual content into the source code buffer of `SourceManager`? How does Clang assign `SourceLocation` to these tokens?

2. When we were talking about implementing a custom `PragmaHandler`, we were leveraging `Preprocessor::Lex` to fetch `Tokens` followed after the pragma name, until we hit the `eod` token kind. Can we keep lexing *beyond* the `eod` token? What interesting things will you do if you can consume arbitrary tokens after the `#pragma` directive?

3. In the `macro guard` project from the *Developing custom preprocessor plugins and callbacks* section, the warning message has the format of `[WARNING] In <source location>: …..` Apparently, this is not a typical compiler warning we see from `clang`, which looks like `<source location>: warning: …`, as shown in the following code snippet:

    ```
    ./simple_warn.c:2:7: warning: unused variable 'y'…
        int y = x + 1;
              ^

      1 warning generated.
    ```

    The `warning` string is even colored in supported terminals. How can we print a warning message such as that? Is there an infrastructure in Clang for doing that?

7
Handling AST

In the previous chapter, we learned how Clang's preprocessor handles preprocessing directives in C-family languages. We also learned how to write different kinds of preprocessor plugins, such as pragma handlers, to extend Clang's functionalities. Those skills are especially useful when it comes to implementing field-specific logic or even custom language features.

In this chapter, we're going to talk about a **semantic-aware** representation of the original source code file once it has been parsed, known as an **Abstract Syntax Tree (AST)**. An AST is a format that carries rich semantic information, including types, expression trees, and symbols, to name a few. It is not only used as a blueprint to generate LLVM IR for later compilation stages but is also the recommended format for performing static analysis. On top of that, Clang also provides a nice framework for developers to intercept and manipulate AST in the middle of the frontend pipeline via a simple plugin interface.

In this chapter, we are going to cover how to process AST in Clang, the important APIs for in-memory AST representation, and how to write AST plugins to implement custom logic with little effort. We will cover the following topics:

- Learning about AST in Clang
- Writing AST plugins

By the end of this chapter, you will know how to work with AST in Clang in order to analyze programs at the source code level. In addition, you will know how to inject custom AST processing logic into Clang in an easy way via AST plugins.

Technical requirements

This chapter expects that you have a build of the clang executable. If you don't, please build it using the following command:

```
$ ninja clang
```

In addition, you can use the following command-line flag to print out the textual representation of AST:

```
$ clang -Xclang -ast-dump foo.c
```

For example, let's say foo.c contains the following content:

```
int foo(int c) { return c + 1; }
```

By using the -Xclang -ast-dump command-line flag, we can print out AST for foo.c:

```
TranslationUnitDecl 0x560f3929f5a8 <<invalid sloc>> <invalid
sloc>
|...
`-FunctionDecl 0x560f392e1350 <foo.c:2:1, col:30> col:5 foo
'int (int)'
  |-ParmVarDecl 0x560f392e1280 <col:9, col:13> col:13 used c
'int'
  `-CompoundStmt 0x560f392e14c8 <col:16, col:30>
    `-ReturnStmt 0x560f392e14b8 <col:17, col:28>
      `-BinaryOperator 0x560f392e1498 <col:24, col:28> 'int'
'+'
        |-ImplicitCastExpr 0x560f392e1480 <col:24> 'int'
<LValueToRValue>
        | `-DeclRefExpr 0x560f392e1440 <col:24> 'int' lvalue
ParmVar 0x560f392e1280 'c' 'int'
        `-IntegerLiteral 0x560f392e1460 <col:28> 'int' 1
```

This flag is useful for finding out what C++ class is used to represent a certain part of the code. For example, the formal function parameter/argument is represented by the `ParmVarDecl` class, which is highlighted in the previous code.

The code samples for this chapter can be found here: `https://github.com/PacktPublishing/LLVM-Techniques-Tips-and-Best-Practices-Clang-and-Middle-End-Libraries/tree/main/Chapter07`.

Learning about AST in Clang

In this section, we are going to learn about Clang's AST in-memory representation and its essential API usage. The first part of this section will provide you with a high-level overview of Clang AST's hierarchy; the second part will focus on a more specific topic regarding type representation in Clang AST; and the final part will show you the basic usage of AST matcher, which is extremely useful when you're writing an AST plugin.

In-memory structure of Clang AST

The in-memory representation of AST in Clang is organized in a hierarchy structure that resembles the syntax structure of C-family language programs. Starting from the top-most level, there are two classes worth mentioning:

- `TranslationUnitDecl`: This class represents an input source file, also called a translation unit (most of the time). It contains all the top-level declarations – global variables, classes, and functions, to name a few – as its children, where each of those top-level declarations has its own subtree that recursively defines the rest of the AST.

- `ASTContext`: As its name suggests, this class keeps track of all the AST nodes and other metadata from the input source files. If there are multiple input source files, each of them gets its own `TranslationUnitDecl`, but they all share the same `ASTContext`.

In addition to the structure, the body of the AST – the AST nodes – can be further classified into three primary categories: **declaration**, **statement**, and **expression**. The nodes in these categories are represented by subclasses derived from the `Decl`, `Expr`, and `Stmt` classes, respectively. In the following sections, we are going to introduce each of these in-memory AST representations.

Declarations

Language constructs such as variable declarations (global and local), functions, and struct/class declarations are represented by subclasses of `Decl`. Though we are not going to go into each of these subclasses here, the following diagram shows common declaration constructions in C/C++ and their corresponding AST classes:

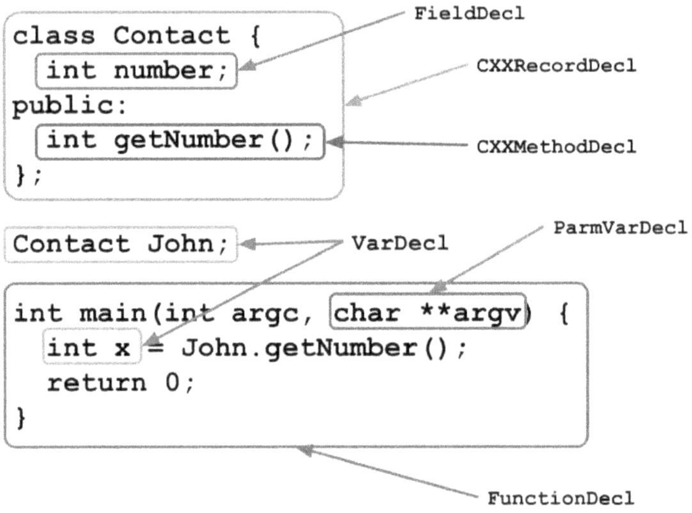

Figure 7.1 – Common declarations in C/C++ and their AST classes

Between more concrete subclasses, such as `FunctionDecl` and `Decl`, there are several important *abstract* classes that represent certain language concepts:

- `NamedDecl`: For every declaration that has a name.

- `ValueDecl`: For declarations whose declared instances can be a value, and thus are associated with type information.

- `DeclaratorDecl`: For every declaration that uses declarator (basically a statement in the form of `<type and qualifier> <identifier name>`). They provide extra information about parts other than the identifier. For example, they provide access to an in-memory object with namespace resolution, which acts as a qualifier in the declarator.

To learn more about AST classes for other kinds of declarations, you can always navigate through the subclasses of `Decl` on LLVM's official API reference website.

Statements

Most directives in a program that represent the concept of *actions* can be classified as statements and are represented by subclasses of `Stmt`, including *expressions*, which we are going to cover shortly. In addition to imperative statements such as function calls or return sites, `Stmt` also covers structural concepts such as `for` loops and `if` statements. Here is a diagram showing a common language construct represented by `Stmt` (except expression) in C/C++ and its corresponding AST classes:

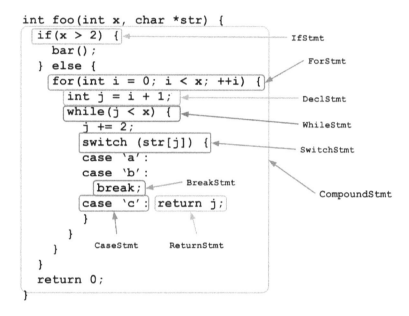

Figure 7.2 – Common statements (excluding expressions) in C/C++ and their AST classes

There are two things worth mentioning about the previous diagram:

- `CompoundStmt`, which is a container for multiple statements, represents not only the function body but basically any code block enclosed by curly braces (`'{'`, `'}'`). Therefore, though not shown in the preceding diagram due to a lack of space, `IfStmt`, `ForStmt`, `WhileStmt`, and `SwitchStmt` all have a `CompoundStmt` child node representing their bodies.

- Declarations in a `CompoundStmt` will be wrapped by a `DeclStmt` node, in which the real `Decl` instance is its child node. This creates a simpler AST design.

Statements are one of the most prevailing directives in a typical C/C++ program. It is worth noting, however, that many statements are organized in a hierarchy (for example, `ForStmt` and its loop body), so it might take you extra steps to go down this hierarchy before you find the desired `Stmt` node.

Expressions

Expressions in Clang AST are a special kind of statement. Different from other statements, expressions always generate *values*. For example, a simple arithmetic expression, *3 + 4*, is expected to generate an integer value. All expressions in Clang AST are represented by subclasses of `Expr`. Here is a diagram showing a common language construct represented by `Expr` in C/C++ and its corresponding AST classes:

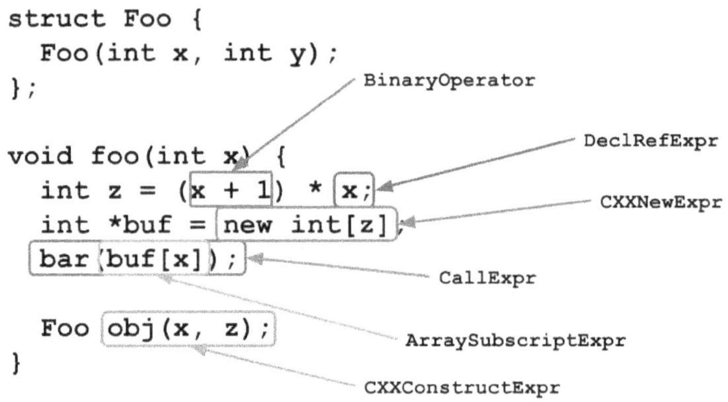

```
struct Foo {
  Foo(int x, int y);
};                              BinaryOperator

                                                    DeclRefExpr
void foo(int x) {
  int z = (x + 1) * x;
                                                    CXXNewExpr
  int *buf = new int[z];
  bar(buf[x]);
                                CallExpr

  Foo obj(x, z);
}                                       ArraySubscriptExpr

                                CXXConstructExpr
```

Figure 7.3 – Common expressions in C/C++ and their AST classes

One important `Expr` class is `DeclRefExpr`. It represents the concept of symbol reference. You can use one of its APIs, `DeclRefExpr::getDecl()`, to retrieve the referenced symbol's `Decl` object. Handy symbol information like this only appears after AST has been generated, so this is one of the reasons people always recommend implementing static analysis logic on AST rather on more primitive forms (inside the parser, for example).

Another interesting `Expr` class – not highlighted in the preceding diagram due to a lack of space – is `ParenExpr`, which represents the parentheses that wrap around an expression. For example, in the preceding diagram, **(x + 1)** is a `ParenExpr` with a `BinaryOperator` representing **x + 1** as its child.

Types in Clang AST

The type system is one of the most crucial components in modern compilers, especially for statically typed languages such as C/C++. Type checking ensures that the input source code is well-formed (to some extent) and catches as many errors as possible at compile time. While we don't need to do type checking by ourselves in Clang, it is done by the `Sema` subsystem, which we introduced in *Chapter 5, Exploring Clang's Architecture*. You will probably need to leverage this information when you're processing the AST. Let's learn how types are modeled in Clang AST.

Core classes

The core of Clang AST's type system is the `clang::Type` class. Each type in the input code – including primitive types such as `int` and user-defined types such as struct/class – is represented by a **singleton type** (more specifically, a *subclass* of `Type`) object.

> **Terminology**
>
> In the rest of this chapter, we will call types in the input source code **source code types**.

A **singleton** is a design pattern that forces a resource or an abstract concept to be represented by only one in-memory object. In our case, source code types are the resources, so you will find only one `Type` object for each of those types. One of the biggest advantages of this design is that you have an easier way to compare two `Type` objects. Let's say you have two `Type` pointers. By doing a simple pointer comparison (which is extremely fast) on them, you can tell if they're representing the same source code type.

> **Counter Example of a Singleton Design**
>
> If `Type` in Clang AST is not using a singleton design, to compare if two `Type` pointers are representing the same source code types, you need to inspect the content of the objects they are pointing to, which is not efficient.

As we mentioned earlier, each source code type is actually represented by a subclass of `Type`. Here are some common `Type` subclasses:

- `BuiltinType`: For primitive types such as `int`, `char`, and `float`.

- `PointerType`: For all the pointer types. It has a function called `PointerType::getPointee()` for retrieving the source code type being pointed to by it.

- `ArrayType`: For all the array types. Note that it has other subclasses for more specific arrays that have either a constant or variable length.

- `RecordType`: For struct/class/union types. It has a function called `RecordType::getDecl()` for retrieving the underlying `RecordDecl`.

- `FunctionType`: For representing a function's signature; that is, a function's argument types and return type (and other properties, such as its calling convention).

Let us now move on to the qualified types.

Qualified types

One of the most confusing things for people new to Clang's code base is that many places use the QualType class rather than subclasses of Type to represent source code types. QualType stands for **qualified type**. It acts as a wrapper around Type to represent concepts such as const <type>, volatile <type>, and restrict <type>*.

To create a QualType from a Type pointer, you can use the following code:

```
// If `T` is representing 'int'…
QualType toConstVolatileTy(Type *T) {
   return QualType(T, Qualifier::Const | Qualifier::Volatile);
} // Then the returned QualType represents `volatile const int`
```

In this section, we learned about the type system in Clang AST. Let's now move on to ASTMatcher, a syntax to match patterns.

ASTMatcher

When we are dealing with a program's AST – for example, we're checking if there is any suboptimal syntax – searching for specific AST nodes *pattern* is usually the first step, and one of the most common things people do. Using the knowledge we learned in the previous section, we know that this kind of pattern matching can be done by iterating through AST nodes via their in-memory classes APIs. For example, given a FunctionDecl – the AST class of a function – you can use the following code to find out if there is a while loop in its body and if the exit condition of that loop is always a literal Boolean value; that is, true:

```
// `FD` has the type of `const FunctionDecl&`
const auto* Body = dyn_cast<CompoundStmt>(FD.getBody());
for(const auto* S : Body->body()) {
   if(const auto* L = dyn_cast<WhileStmt>(S)) {
      if(const auto* Cond = dyn_cast<CXXBoolLiteralExpr>
        (L->getCond()))
        if(Cond->getValue()) {
          // The exit condition is `true`!!
        }
   }
}
```

As you can see, it created more than three (indention) layers of `if` statements to complete such a simple check. Not to mention in real-world cases, we need to insert even more sanity checks among these lines! While Clang's AST design is not hard to understand, we need a more *concise* syntax to complete pattern matching jobs. Fortunately, Clang has already provided one – the **ASTMatcher**.

ASTMatcher is the utility that helps you write AST pattern matching logic via a clean, concise, and efficient **Domain-Specific Language (DSL)**. Using ASTMatcher, doing the same matching shown in the previous snippet only takes few lines of code:

```
functionDecl(compountStmt(hasAnySubstatement(
  whileStmt(
    hasCondition(cxxBoolLiteral(equals(true)))))));
```

Most of the directives in the preceding snippet are pretty straightforward: function calls such as `compoundStmt(...)` and `whileStmt(...)` check if the current node matches a specific node type. Here, the arguments in these function calls either represent pattern matchers on their subtree or check additional properties of the current node. There are also other directives for expressing qualifying concepts (for example, *for all substatements in this loop body, a return value exists*), such as `hasAnySubstatement(...)`, and directives for expressing data type and constant values such as the combination of `cxxBoolLiteral(equals(true))`.

In short, using ASTMatcher can make your pattern matching logic more *expressive*. In this section, we showed you the basic usage of this elegant DSL.

Traversing AST

Before we dive into the core syntax, let's learn how ASTMatcher traverses AST and how it passes the result back to users after the matching process is completed.

`MatchFinder` is a commonly used driver for the pattern matching process. Its basic usage is pretty simple:

```
using namespace ast_matchers;
...
MatchFinder Finder;
// Add AST matching patterns to `MatchFinder`
Finder.addMatch(traverse(TK_AsIs, pattern1), Callback1);
Finder.addMatch(traverse(TK_AsIs, pattern2), Callback2);
...
// Match a given AST. `Tree` has the type of `ASTContext&`
```

```
// If there is a match in either of the above patterns,
// functions in Callback1 or Callback2 will be invoked
// accordingly
Finder.matchAST(Tree);


// …Or match a specific AST node. `FD` has the type of
// `FunctionDecl&`
Finder.match(FD, Tree);
```

`pattern1` and `pattern2` are pattern objects that are constructed by DSL, as shown previously. What's more interesting is the `traverse` function and the `TK_AsIs` argument. The `traverse` function is a part of the pattern matching DSL, but instead of expressing patterns, it describes the action of traversing AST nodes. On top of that, the `TK_AsIs` argument represents the *traversing mode*.

When we showed you the command-line flag for dumping AST in textual format (`-Xclang -ast-dump`) earlier in this chapter, you may have found that many *hidden AST nodes* were inserted into the tree to assist with the program's semantics rather than representing the real code that was written by the programmers. For example, `ImplicitCastExpr` is inserted in lots of places to ensure the program's type correctness. Dealing with these nodes might be a painful experience when you're composing pattern matching logic. Thus, the `traverse` function provides an alternative, *simplified*, way to traverse the tree. Let's say we have the following input source code:

```
struct B {
  B(int);
};
B foo() { return 87; }
```

When you pass `TK_AsIs` as the first argument to `traverse`, it observes the tree, similar to how `-ast-dump` does:

```
FunctionDecl
`-CompoundStmt
  `-ReturnStmt
    `-ExprWithCleanups
      `-CXXConstructExpr
        `-MaterializeTemporaryExpr
          `-ImplicitCastExpr
            `-ImplicitCastExpr
```

```
`-CXXConstructExpr
  `-IntegerLiteral 'int' 87
```

However, by using `TK_IgnoreUnlessSpelledInSource` as the first argument, the tree that's observed is equal to the following one:

```
FunctionDecl
`-CompoundStmt
  `-ReturnStmt
    `-IntegerLiteral 'int' 87
```

As its name suggests, `TK_IgnoreUnlessSpelledInSource` only visit nodes that are really shown in the source code. This greatly simplifies the process of writing a matching pattern since we don't need to worry about the nitty-gritty details of AST anymore.

On the other hand, `Callback1` and `Callback2` in the first snippet are `MatchFinder::MatchCallback` objects that describe the actions to perform when there is a match. Here is the skeleton of a `MatchCallback` implementation:

```
struct MyMatchCallback : public MatchFinder::MatchCallback {
  void run(const MatchFinder::MatchResult &Result) override {
    // Reach here if there is a match on the corresponding
    // pattern
    // Handling "bound" result from `Result`, if there is any
  }
};
```

In the next section, we will show you how to bind a specific part of the pattern with a tag and retrieve it in `MatchCallback`.

Last but not least, though we used `MatchFinder::match` and `MatchFinder::matchAST` in the first snippet to kick off the matching process, there are other ways to do this. For example, you can use `MatchFinder::newASTConsumer` to create an `ASTConsumer` instance that will run the described pattern matching activity. Alternatively, you can use `ast_matchers::match(...)` (not a member function under `MatchFinder` but a standalone function) to perform matching on a provided pattern and `ASTContext` in a single run, before returning the matched node.

ASTMatcher DSL

ASTMatcher provides an easy-to-use and concise C++ DSL to help with matching AST. As we saw earlier, the *structure* of the desired pattern is expressed by nested function calls, where each of these functions represents the *type* of AST node to match.

Using this DSL to express simple patterns cannot be easier. However, when you're trying to compose patterns with multiple conditions/predicates, things get a little bit more complicated. For example, although we know a for loop (for example, `for(I = 0; I < 10; ++I){…}`) can be matched by the `forStmt(…)` directive, how do we add a condition to its initialize statement (`I = 0`) and exit the condition (`I < 10`) or its loop body? Not only does the official API reference site (the doxygen website we usually use) lacks clear documentation on this part, most of these DSL functions are also pretty flexible in how they accept a wide range of arguments as their subpatterns. For example, following the question on matching a `for` loop, you can use the following code to check only the loop's body:

```
forStmt(hasBody(…));
```

Alternatively, you can check its loop body and its exit condition, like so:

```
forStmt(hasBody(…),
        hasCondition(…));
```

A generalized version of this question would be, given an arbitrary DSL directive, how do we know the *available* directives that can be combined with it?

To answer this question, we will leverage a documentation website LLVM specifically created for ASTMatcher: `https://clang.llvm.org/docs/LibASTMatchersReference.html`. This website consists of a huge three-column table showing the returned type and argument types for each of the DSL directives:

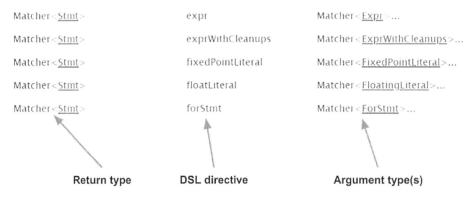

Figure 7.4 – Part of the ASTMatcher DSL reference

Though this table is just a simplified version of normal API references, it already shows you how to search for candidate directives. For example, now that you know forStmt(...) takes zero or multiple Matcher<ForStmt>, we can search this table for directives that return either Matcher<ForStmt> or Matcher<(parent class of ForStmt)>, such as Matcher<Stmt>. In this case, we can quickly spot hasCondition, hasBody, hasIncrement, or hasLoopInit as candidates (of course, many other directives that return Matcher<Stmt> can also be used).

When you're performing pattern matching, there are many cases where you not only want to know if a pattern matches or not but also get the matched AST nodes. In the context of ASTMatcher, its DSL directives only check the *type* of the AST nodes. If you want to retrieve (part of the) concrete AST nodes that are being matched, you can use the bind(...) API. Here is an example:

```
forStmt(
  hasCondition(
    expr().bind("exit_condition")));
```

Here, we used expr() as a wildcard pattern to match any Expr node. This directive also calls bind(...) to associate the matched Expr AST node with the name exit_condition.

Then, in MatchCallback, which we introduced earlier, we can retrieve the bound node by using the following code:

```
...
void run(const MatchFinder::MatchResult &Result) override {
  cons auto& Nodes = Result.Nodes;
  const Expr* CondExpr = Nodes.getNodeAs<Expr>
    ("exit_condition");
  // Use `CondExpr`...
}
```

The getNodeAs<...>(...) function tries to fetch the bound AST node under the given name and cast it to the type suggested by the template argument.

Note that you're allowed to bind different AST nodes under the same name, in which case only the last bounded one will be shown in MatchCallback::run.

Putting everything together

Now that you've learned about both the pattern matching DSL syntax and how to traverse AST using ASTMatcher, let's put these two things together.

Let's say we want to know the number of iterations – also known as the *trip count* – that a simple `for` loop (the loop index starts from zero and is incremented by one at each iteration and bounded by a literal integer) has in a function:

1. First, we must write the following code for matching and traversing:

```
auto PatExitCondition = binaryOperator(
                          hasOperatorName("<"),
                          hasRHS(integerLiteral()
                          .bind("trip_count")));
auto Pattern = functionDecl(
                  compoundStmt(hasAnySubstatement(
              forStmt(hasCondition(PatExitCondition)))));

MatchFinder Finder;
auto* Callback = new MyMatchCallback();
Finder.addMatcher(traverse(TK_IgnoreUnlessSpelledInSource,
                            Pattern), Callback);
```

The preceding snippet also shows how *modular* DSL patterns are. You can create individual pattern fragments and compose them depending on your needs, as long as they're compatible.

Finally, here is what `MyMatchCallback::run` looks like:

```
void run(const MatchFinder::MatchResult &Result) override
{
  const auto& Nodes = Result.Nodes;
  const auto* TripCount =
        Nodes.getNodeAs<IntegerLiteral>("trip_count");
  if (TripCount)
    TripCount->dump(); // print to llvm::errs()
}
```

2. After this, you can use `Finder` to match the desired pattern (either by calling `MatchFinder::match` or `MatchFinder::matchAST`, or by creating an `ASTConsumer` using `MatchFinder::newASTConsumer`) on an AST. The matched trip count will be printed to `stderr`. For instance, if the input source code is `for(int i = 0; i < 10; ++i) {...}`, the output will simply be `10`.

In this section, we learned how Clang structures its AST, how Clang AST is represented in memory, and how to use ASTMatcher to help developers with AST pattern matching. With this knowledge, in the next section, we will show you how to create an AST plugin, which is one of the easiest ways to inject custom logic into Clang's compilation pipeline.

Writing AST plugins

In the previous section, we learned how AST is represented in Clang and learned what its in-memory classes look like. We also learned about some useful skills we can use to perform pattern matching on Clang AST. In this section, we will learn how to write plugins that allow you to insert custom AST processing logic into Clang's compilation pipeline.

This section will be divided into three parts:

- **Project overview**: The goal and overview of the demo project we are going to create in this section.

- **Printing diagnostic messages**: Before we dive into the core of developing a plugin, we are going to learn how to use Clang's `DiagnosticsEngine`, a powerful subsystem that helps you print out well-formatted and meaningful diagnostic messages. This will make our demo project more applicable to real-world scenarios.

- **Creating the AST plugin**: This section will show you how to create an AST plugin from scratch, fill in all the implementation details, and how to run it with Clang.

Project overview

In this section, we will create a plugin that prompts the user with warning messages whenever there are `if-else` statements in the input code that can be converted into **ternary operators**.

Quick Refresher – Ternary Operator

The ternary operator, `x? val_1 : val_2`, is evaluated to `val_1` when the `x` condition is true. Otherwise, it is evaluated to `val_2`.

For example, let's look at the following C/C++ snippet:

```
int foo(int c) {
  if (c > 10) {
    return c + 100;
  } else {
    return 94;
  }
}

void bar(int x) {
  int a;
  if (x > 10) {
    a = 87;
  } else {
    a = x - 100;
  }
}
```

The if-else statements in both functions can be converted into ternary operators, like this:

```
int foo(int c) {
  return c > 10? c + 100 : 94;
}

void bar(int x) {
  int a;
  a = x > 10? 87 : x - 100;
}
```

In this project, we will only focus on finding two kinds of potential ternary operator opportunities:

- Both the then block (true branch) and the else block (false branch) contain a single return statement. In this case, we can coalesce their return values and the branch condition into one ternary operator (as the new returned value).

- Both the `then` block and the `else` block only contain a single assignment statement. Both statements use a single `DeclRefExpr` – that is, a symbol reference – as the LHS, and both `DeclRefExpr` objects point to the same `Decl` (symbol). In other words, we are covering the case of the `bar` function shown in the preceding snippet. Note that we are not covering cases where the LHS is more complicated; for example, where an array subscription, `a[i]`, is used as the LHS.

After identifying these patterns, we must prompt warning messages to the user and provide extra information to help the user fix this issue:

```
$ clang …(flags to run the plugin) ./test.c
./test.c:2:3: warning: this if statement can be converted to
ternary operator:
  if (c > 10) {
  ^
./test.c:3:12: note: with true expression being this:
    return c + 100;
           ^
./test.c:5:12: note: with false expression being this:
    return 94;
           ^
./test.c:11:3: warning: this if statement can be converted to
ternary operator:
  if (x > 10) {
  ^
./test.c:12:9: note: with true expression being this:
    a = 87;
        ^
./test.c:14:9: note: with false expression being this:
    a = x - 100;
        ^
2 warnings generated.
```

Each warning message – which tells you which `if-else` statement can be converted into a ternary operator – is followed by two notes pointing out the potential expressions to construct for the operator.

Compared to handcrafting compiler messages, as we did in the *Developing custom preprocessor plugins and callbacks* section of *Chapter 6, Extending the Preprocessor*, here, we are using Clang's diagnostics infrastructure to print messages that carry richer information, such as the snapshot of code that the message is referring to. We will show you how to use that diagnostic infrastructure next.

Printing diagnostic messages

In *Chapter 6, Extending the Preprocessor*, we asked if you could improve the warning message format in the example project shown in the *Developing custom preprocessor plugins and callbacks* section, so that it's closer to the compiler messages you saw from Clang. One of the solutions to that question is using Clang's diagnostic framework. We are going to look at this in this section.

Clang's diagnostic framework consists of three primary parts:

- **Diagnostic IDs**
- **Diagnostic engine**
- **Diagnostic consumers (clients)**

Their relationships can be seen in the following diagram:

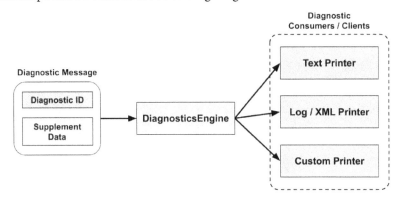

Figure 7.5 – High-level organization of Clang's diagnostic framework

Diagnostic messages

Starting from the left-hand side of the preceding diagram, most of the time, a diagnostic message – for example, *use of undeclared identifier "x"* – is associated with a message **template** that has its own diagnostic ID. Using the undeclared identifier message, for example, its message template looks like this:

```
"use of undeclared identifier %0"
```

%0 is a **placeholder** that will be filled in by supplemental data later. In this case, it is the concrete identifier name (x, in the preceding example message). The number following % also suggests which supplemental data it will use. We will cover this format in detail shortly.

Templates are registered with the diagnostic engine via TableGen syntax. For example, the message we are discussing here is put inside clang/include/clang/Basic/ DiagnosticSemaKinds.td:

```
def err_undeclared_var_use : Error<"use of undeclared
identifier %0">;
```

We highlighted two parts in the preceding snippet. First, the name of this message template, err_undeclared_var_use, will be used later as the unique diagnostic ID. Second, the Error TableGen class suggested that this is an error message, or more formally speaking, its *diagnostic level* error.

In summary, a diagnostic message consists of a unique diagnostic ID – which is associated with a message template and its diagnostic level – and the supplemental data to put in the placeholders of the template, if there are any.

Diagnostic consumers

After the diagnostic message is sent to the diagnostic engine – represented by the DiagnosticsEngine class – the engine formats the messages into textual contents and send them to one of the **diagnostic consumers** (also called **clients** in the code base; we will use the term **consumer** in rest of this section).

A diagnostic consumer – an implementation of the DiagnosticConsumer class – post-processes the textual messages sent from DiagnosticsEngine and exports them via different mediums. For example, the default TextDiagnosticPrinter prints messages to the command-line interface; LogDiagnosticPrinter, on the other hand, decorates the incoming messages with simple XML tags before printing them into log files. In theory, you can even create a custom DiagnosticConsumer that sends diagnostic messages to a remote host!

Reporting diagnostic messages

Now that you have learned how Clang's diagnostic framework works, let's learn how to send (report) a diagnostic message to `DiagnosticEngine`:

1. First, we need to retrieve a reference to `DiagnosticEngine`. The engine itself is sitting at the core of Clang's compilation pipeline, so you can fetch it from various primary components, such as `ASTContext` and `SourceManager`. The following is an example:

    ```
    // `Ctx` has the type of `ASTContext&`
    DiagnosticsEngine& Diag = Ctx.getDiagnostics();
    ```

2. Next, we need to use the `DiagnosticsEngine::Report` function. This function always takes a diagnostic ID as one of its arguments. For example, to report `err_undeclared_var_use`, which we introduced earlier, use the following code:

    ```
    Diag.Report(diag::err_undeclared_var_use);
    ```

    However, we know that `err_undeclared_var_use` takes one placeholder argument – namely, the identifier name – which is supplied through concatenating the `Report` function call with `<<` operators:

    ```
    Diag.Report(diag::err_undeclared_var_use) << ident_name_
    str;
    ```

3. Recall that `err_undeclared_var_use` only has one placeholder, `%0`, so it picks up the first values in the following `<<` stream. Let's say we have a diagnostic message, `err_invalid_placement`, with the following template:

    ```
    "you cannot put %1 into %0"
    ```

4. You can report this using the following code:

    ```
    Diag.Report(diag::err_invalid_placement)
                  << "boiling oil" << "water";
    ```

5. In addition to simple placeholders, another useful feature is the `%select` directive. For example, we have a diagnostic message, `warn_exceed_limit`, with a template like this:

    ```
    "you exceed the daily %select{wifi|cellular network}0
    limit"
    ```

The `%select` directive consists of curly braces in which different message options are separated by `|`. Outside the curly braces, a number – 0, in the preceding code – indicates which supplement data is used to select the option within the braces. The following is an example of this:

```
Diag.Report(diag::warn_exceed_limit) << 1;
```

The preceding snippet will output **You exceed the daily cellular network limit**. Let's say you use 0 as the parameter after the stream operator (`<<`):

```
Diag.Report(diag::warn_exceed_limit) << 0;
```

This will result in a message stating **you exceed the daily wifi limit**.

6. Now, let's say you use another version of the Report function, which takes an additional `SourceLocation` argument:

```
// `SLoc` has the type of `SourceLocation`
Diag.Report(SLoc, diag::err_undeclared_var_use)
                    << ident_name_str;
```

The output message will contain part of the source code being pointed to by SLoc:

```
test.cc:2:10: error: use of undeclared identifier 'x'
    return x + 1;
           ^
```

7. Last but not least, though most of the diagnostic messages are registered with `DiagnosticsEngine` via TableGen code put *inside* Clang's source tree, this doesn't mean that developers cannot create their new diagnostic messages without modifying Clang's source tree. Let's introduce `DiagnosticsEngine::getCustomDiagID(…)`, the API that creates a new diagnostic ID from a message template and diagnostic level provided by developers:

```
auto MyDiagID = Diag.
getCustomDiagID(DiagnosticsEngine::Note,
"Today's weather is %0");
```

The preceding snippet creates a new diagnostic ID, `MyDiagID`, that has a message template of **Today's weather is %0** at its note diagnostic level. You can use this diagnostic ID just like any other ID:

```
Diag.Report(MyDiagID) << "cloudy";
```

In this section, you learned how to leverage Clang's diagnostic framework to print out messages just like normal compiler messages.

Next, we are going to combine all the skills we've learned about in this chapter to create a custom AST plugin.

Creating the AST plugin

In the previous sections of this chapter, we explored Clang's AST and learned how to use it in in-memory APIs. In this section, we will learn how to write a plugin that helps you insert your custom AST processing logic into Clang's compilation pipeline in an easy way.

In *Chapter 5, Exploring Clang's Architecture*, we learned about the advantages of using Clang (AST) plugins: they can be developed even you are using a prebuilt `clang` executable, they are easy to write, and they have good integration with the existing toolchain and build systems, to name a few. In *Chapter 6, Extending the Preprocessor*, we developed a plugin for custom pragma handling in the preprocessor. In this chapter, we will also be writing a plugin, but this one will be designed for custom AST processing. The code skeletons for these two plugins are also quite different.

We introduced the sample project we will be using in this section in the *Project overview* section. This plugin will prompt users with warning messages if some `if-else` statements in the input code can be converted into ternary operators. In addition, it also shows extra hints about candidate expressions for building the ternary operator.

Here are the detailed steps for building the plugin:

1. Similar to the pragma plugin we saw in *Chapter 6, Extending the Preprocessor*, creating a plugin in Clang is basically like implementing a class. In the case of the AST plugin, this will be the `PluginASTAction` class.

    `PluginASTAction` is a subclass of `ASTFrontendAction` – a `FrontendAction` specialized for handling AST (if you're not familiar with `FrontendAction`, feel free to read *Chapter 5, Exploring Clang's Architecture*, again). Thus, we need to implement the `CreateASTConsumer` member function:

    ```
    struct TernaryConverterAction : public PluginASTAction {
      std::unique_ptr<ASTConsumer>
        CreateASTConsumer(CompilerInstance &CI,
                          StringRef InFile) override;
    };
    ```

    We will fill in this function later.

2. In addition to `CreateASTConsumer`, there are two other member functions we can override to change some of the functionalities: `getActionType` and `ParseArgs`. The former tells Clang *how* this plugin should be executed by returning one of the enum values shown here:

    a. `Cmdline`: The plugin will be executed after the main action if users provide the `-plugin <plugin name>` (frontend) command-line flag.

    b. `ReplaceAction`: This replaces the original action Clang was going to perform. For example, if Clang was supposed to compile input code into an object file (the `-c` flag), it will execute the plugin's action instead once the plugin has been loaded.

    c. `AddBefore/AfterMainAction`: The original Clang action will still be executed, and the plugin action will be prepended/appended to it.

    Here, we will use the `Cmdline` action type:

    ```
    struct TernaryConverterAction : public PluginASTAction {
      ...
      ActionType getActionType() override { return Cmdline; }
    };
    ```

    The `ParseArgs` member function, on the other hand, handles (frontend) command-line options specific to this plugin. In other words, you can create custom command-line flags for your plugin. In our case, we are going to create two flags: `-no-detect-return` and `-no-detect-assignment`. This allows us to decide whether we wish to detect potential ternary conversions regarding `return` statements or assignment statements, respectively:

    ```
    struct TernaryConverterAction : public PluginASTAction {
      ...
      bool NoAssignment = false,
           NoReturn = false;

      bool ParseArgs(const CompilerInstance &CI,
            const std::vector<std::string> &Args) override {
        for (const auto &Arg : Args) {
          if (Arg == "-no-detect-assignment") NoAssignment =
            true;
          if (Arg == "-no-detect-return") NoReturn = true;
        }
    ```

```
    return true;
  }
};
```

As shown in the preceding snippet, we created two boolean flags, `NoReturn` and `NoAssignment`, to carry our command-line options' values. An important thing to know is the return value for `ParseArgs`. Instead of returning *if it parsed any custom flag*, `ParseArgs` is actually returning *if the plugin should continue its execution*. Therefore, you should always return true in most cases.

3. Now, we are going to talk about the content of `CreateASTConsumer`. This function will return an `ASTConsumer` object, which is the main body that we will put our custom logic in. Nevertheless, we are not going to directly implement an `ASTConsumer`. Instead, we are going to us the `ASTConsumer` object that was generated by *ASTMatcher*, which we introduced earlier in this chapter.

Recall that two things are required to build a `MatchFinder` instance – the primary pattern matching driver in ASTMatcher (patterns written in ASTMatcher's own DSL) and a `MatchCallback` implementation. Let's separate our patterns and matcher callbacks into two categories: patterns for detecting potential ternary operator opportunities based on `return` statements and those for detecting *assignment-statement-based opportunities*.

Here is the skeleton for `CreateASTConsumer`:

```
using namespace ast_matchers;
struct TernaryConverterAction : public PluginASTAction {
  ...
private:
  std::unique_ptr<MatchFinder> ASTFinder;
  std::unique_ptr<MatchFinder::MatchCallback>
    ReturnMatchCB, AssignMatchCB;
};

std::unique_ptr<ASTConsumer>
TernaryConverterAction::CreateASTConsumer
(CompilerInstance &CI, StringRef InFile) {
  ASTFinder = std::make_unique<MatchFinder>();
  // Return matcher
  if (!NoReturn) {
    ReturnMatchCB = /*TODO: Build MatcherCallback
```

```
      instance*/
   ASTFinder->addMatcher(traverse
     (TK_IgnoreUnlessSpelledInSource,
       /*TODO: Patterns in DSL*/), ReturnMatchCB.get());
   }
   // Assignment matcher
   if (!NoAssignment) {
     AssignMatchCB = /*TODO: Build MatcherCallback
       instance*/
     ASTFinder->addMatcher(traverse
       (TK_IgnoreUnlessSpelledInSource,
         /*TODO: Patterns in DSL*/), AssignMatchCB.get());
   }
   return std::move(ASTFinder->newASTConsumer());
}
```

The preceding code created three additional `unique_ptr` type member variables: one for holding `MatchFinder` and two `MatchCallback` ones for return-based and assignment-based patterns.

Why Use unique_ptr?

The rationale behind using `unique_ptr` to store those three objects – or storing those objects *persistently* – is because the `ASTConsumer` instance we created at the end of `CreateASTConsumer` (`ASTFinder->newASTConsumer()`) keeps references to those three objects. Thus, we need a way to keep them alive during the lifetime of the frontend.

In addition to that, we registered the pattern for traversal with MatchFinder by using `MatchFinder::addMatcher`, the `traverse` function, and `MatchCallback` instances. If you're not familiar with these APIs, feel free to check out the *ASTMatcher* section.

Now, we only need to compose the matching patterns and implement some callbacks to print out warning messages if there is a match – as the TODO comments suggested in the preceding snippet.

4. Let's deal with the patterns first. The patterns we are looking for – both return-based and assignment-based patterns – have if-else statements (IfStmt) enclosed by a function (FunctionDecl for the entire function and CompoundStmt for the function body) in their outermost layout. Inside both, in the true branch and false branch of IfStmt, only one statement can exist. This structure can be illustrated like so:

```
FunctionDecl
  |_CompoundStmt
     |_(Other AST nodes we don't care)
     |_IfStmt
        |_(true branch: contain only one return/assign
          statement)
        |_(false branch: contain only one return/assign
          statement)
```

To convert this concept into ASTMatcher's DSL, here is the DSL code that's shared between the return-based and assignment-based patterns:

```
functionDecl(
  compoundStmt(hasAnySubstatement
    IfStmt(
      hasThen(/*TODO: Sub-pattern*/)
      hasElse(/*TODO: Sub-pattern*/)
    )
  )
);
```

One important thing to remember is that when you're dealing with CompoundStmt, you should always use quantifier directives such as hasAnySubstatement to match its body statements.

We are going to use the previous TODO comments to customize for either return-based or assignment-based situations. Let's use subpattern variables to replace those TODO comments and put the preceding code into another function:

```
StatementMatcher
buildIfStmtMatcher(StatementMatcher truePattern,
```

```
                    StatementMatcher falsePattern) {
  return functionDecl(
    compoundStmt(hasAnySubstatement
      IfStmt(
        hasThen(truePattern)
        hasElse(falsePattern))));
}
```

5. For return-based patterns, the subpatterns for both the `if-else` branches
    mentioned in the previous step are identical and simple. We're also using a separate
    function to create this pattern:

```
StatementMatcher buildReturnMatcher() {
  return compoundStmt(statementCountIs(1),
                      hasAnySubstatement(
                      returnStmt(
                        hasReturnValue(expr())))));
}
```

As shown in the preceding snippet, we are using the `statementCountIs`
directive to match the code blocks with only one statement. Also, we specified that
we don't want an empty return via `hasReturnValue(...)`. The argument for
`hasReturnValue` is necessary since the latter takes at least one argument, but
since we don't care what type of node it is, we are using `expr()` as some sort of
wildcard pattern.

For assignment-based patterns, things get a little bit complicated: we don't just
want to match a single assignment statement (modeled by the `BinaryOperator`
class) in both branches – the LHS of those assignments need to be `DeclRefExpr`
expressions that point to the same `Decl` instance. Unfortunately, we are not able
to express all these predicates using ASTMatch's DSL. What we can do, however,
is push off some of those checks into `MatchCallback` later, and only use DSL
directives to check the *shape* of our desired patterns:

```
StatementMatcher buildAssignmentMatcher() {
  return compoundStmt(statementCountIs(1),
                      hasAnySubstatement(
                      binaryOperator(
                        hasOperatorName("="),
                        hasLHS(declRefExpr())
```

```
                                  )));
    }
```

6. Now that we've completed the skeleton for our patterns, it's time to implement `MatchCallback`. There are two things we are going to do in `MatchCallback::run`. First, for our assignment-based pattern, we need to check if the LHS' `DeclRefExpr` of those matched assignment candidates is pointing to the same `Decl`. Second, we want to print out messages that help users rewrite `if-else` branches as ternary operators. In other words, we need location information from some of the matched AST nodes.

   Let's solve the first task using the *AST node binding technique*. The plan is to bind the candidate assignment's LHS `DeclRefExpr` nodes so that we can retrieve them from `MatchCallback::run` later and perform further checks on their `Decl` nodes. Let's change `buildAssignmentMatch` into this:

   ```
   StatementMatcher buildAssignmentMatcher() {
       return compoundStmt(statementCountIs(1),
                           hasAnySubstatement(
                               binaryOperator(
                                   hasOperatorName("="),
                                   hasLHS(declRefExpr().
                                   bind("dest")))));
   }
   ```

   Though the preceding code seems straightforward, there is one problem in this binding scheme: in both branches, `DeclRefExpr` is bound to the same name, meaning that the AST node that occurred later will overwrite the previously bound node. So, eventually, we won't get `DeclRefExpr` nodes from both branches as we previously planned.

   Therefore, let's use a different tags for `DeclRefExpr` that match from both branches: `dest.true` for the true branch and `dest.false` for the false branch. Let's tweak the preceding code to reflect this strategy:

   ```
   StatementMatcher buildAssignmentMatcher(StringRef Suffix)
   {
       auto DestTag = ("dest" + Suffix).str();
       return compoundStmt(statementCountIs(1),
                           hasAnySubstatement(
                               binaryOperator(
   ```

```
                hasOperatorName("="),
                hasLHS(declRefExpr().
                  bind(DestTag))))));
}
```

Later, when we call `buildAssignmentMatcher`, we will pass different suffixes for the different branches – either `.true` or `.false`.

Finally, we must retrieve the bound nodes in `MatchCallback::run`. Here, we are creating different `MatchCallback` subclasses for return-based and assignment-based scenarios – `MatchReturnCallback` and `MatchAssignmentCallback`, respectively. Here is a part of the code in `MatchAssignmentCallback::run`:

```
void
MatchAssignmentCallback::run(const MatchResult &Result)
override {
  const auto& Nodes = Result.Nodes;
  // Check if destination of both assignments are the
  // same
  const auto *DestTrue =
            Nodes.getNodeAs<DeclRefExpr>("dest.true"),
            *DestFalse =
            Nodes.getNodeAs<DeclRefExpr>("dest.false");
  if (DestTrue->getDecl() == DestFalse->getDecl()) {
    // Can be converted into ternary operator!
  }
}
```

We are going to solve the second task – printing useful information to users – in the next step.

7. To print useful information – including *which* part of the code can be converted into a ternary operator, and *how* can you build that ternary operator – we need to retrieve some AST nodes from the matched patterns before getting their source location information. For this, we will use some node binding tricks, as we did in the previous step. This time, we will modify all the pattern building functions; that is, `buildIfStmtMatcher`, `buildReturnMatcher`, and `buildAssignmentMatcher`:

```
StatementMatcher
buildIfStmtMatcher(StatementMatcher truePattern,
                    StatementMatcher falsePattern) {
  return functionDecl(
    compoundStmt(hasAnySubstatement
      IfStmt(
        hasThen(truePattern)
        hasElse(falsePattern)).bind("if_stmt")
    ));
}
```

Here, we bound the matched `IfStmt` since we want to tell our users where the potential places that can be converted into ternary operators are:

```
StatementMatcher buildReturnMatcher(StringRef Suffix) {
  auto Tag = ("return" + Suffix).str();
  return compoundStmt(statementCountIs(1),
                        hasAnySubstatement(
                          returnStmt(hasReturnValue(
                            expr().bind(Tag)
                          )))));
}
StatementMatcher buildAssignmentMatcher(StringRef Suffix)
{
  auto DestTag = ("dest" + Suffix).str();
  auto ValTag = ("val" + Suffix).str();
  return compoundStmt(statementCountIs(1),
                        hasAnySubstatement(
                          binaryOperator(
                            hasOperatorName("="),
                            hasLHS(declRefExpr().
```

```
                    bind(DestTag)),
                 hasRHS(expr().bind(ValTag))
               )));
}
```

We used the same node binding tricks here that we did in the preceding snippet. After this, we can retrieve those bound nodes from `MatchCallback::run` and print out the message using the `SourceLocation` information that's attached to those nodes.

We are going to use Clang's diagnostic framework to print out those messages here (feel free to read the *Printing diagnostic messages* section again if you're not familiar with it). And since the prospective message formats are not existing ones in Clang's code base, we are going to create our own diagnostic ID via `DiagnosticsEngine::getCustomDiagID(...)`. Here is what we will do in `MatchAssignmentCallback::run` (we will only demo `MatchAssignmentCallback` here since `MatchReturnCallback` is similar):

```
void
MatchAssignmentCallback::run(const MatchResult &Result)
override {
  ...
  auto& Diag = Result.Context->getDiagnostics();
  auto DiagWarnMain = Diag.getCustomDiagID(
    DiagnosticsEngine::Warning,
    "this if statement can be converted to ternary
     operator:");

  auto DiagNoteTrueExpr = Diag.getCustomDiagID(
    DiagnosticsEngine::Note,
    "with true expression being this:");

  auto DiagNoteFalseExpr = Diag.getCustomDiagID(
    DiagnosticsEngine::Note,
    "with false expression being this:");
  ...
}
```

Combining this with bound node retrievals, here is how we are going to print the messages:

```
void
MatchAssignmentCallback::run(const MatchResult &Result)
override {

    ...

    if (DestTrue && DestFalse) {
        if (DestTrue->getDecl() == DestFalse->getDecl()) {
            // Can be converted to ternary!
            const auto* If = Nodes.getNodeAs<IfStmt>
            ("if_stmt");
            Diag.Report(If->getBeginLoc(), DiagWarnMain);

            const auto* TrueValExpr =
                        Nodes.getNodeAs<Expr>("val.true");
            const auto* FalseValExpr =
                        Nodes.getNodeAs<Expr>("val.false");
            Diag.Report(TrueValExpr->getBeginLoc(),
                        DiagNoteTrueExpr);
            Diag.Report(FalseValExpr->getBeginLoc(),
                        DiagNoteFalseExpr);
        }
    }
}
```

8. Finally, go back to `CreateASTConsumer`. Here is how everything is pieced together:

```
std::unique_ptr<ASTConsumer>
TernaryConverterAction::CreateASTConsumer(
CompilerInstance &CI, StringRef InFile) {

    ...

    // Return matcher
    if (!NoReturn) {
    ReturnMatchCB = std::make_unique<MatchReturnCallback>();
        ASTFinder->addMatcher(
            traverse(TK_IgnoreUnlessSpelledInSource,
                    buildIfStmtMatcher(
```

```
                    buildReturnMatcher(".true"),
                    buildReturnMatcher(".false"))),
      ReturnMatchCB.get()
  );
}


  // Assignment matcher
  if (!NoAssignment) {
    AssignMatchCB = std::make_
      unique<MatchAssignmentCallback>();
    ASTFinder->addMatcher(
      traverse(TK_IgnoreUnlessSpelledInSource,
              buildIfStmtMatcher(
                buildAssignmentMatcher(".true"),
                buildAssignmentMatcher(".false"))),
      AssignMatchCB.get()
    );
  }

    return std::move(ASTFinder->newASTConsumer());
}
```

And that wraps up all the things we need to do!

9. Last but not least, this is the command for running our plugin:

```
$ clang -fsyntax-only -fplugin=/path/to/TernaryConverter.
so -Xclang -plugin -Xclang ternary-converter \
    test.c
```

You will get an output similar to the one you saw in the *Project overview* section.

To use plugin-specific flags, such as `-no-detect-return` and `-no-detect-assignment` in this project, please add the command-line options highlighted here:

```
$ clang -fsyntax-only -fplugin=/path/to/TernaryConverter.
so -Xclang -plugin -Xclang ternary-converter \
    -Xclang -plugin-arg-ternary-converter \
    -Xclang -no-detect-return \
    test.c
```

To be more specific, the first highlighted argument is in `-plugin-arg-<plugin name>` format.

In this section, you learned how to write an AST plugin that sends messages to users whenever there is an `if-else` statement that can be converted into a ternary operator. You did this by leveraging all the techniques that were covered in this chapter; that is, Clang AST's in-memory representation, ASTMatcher, and the diagnostic framework, to name a few.

Summary

When it comes to program analysis, AST is usually the recommended medium to use, thanks to its rich amount of semantic information and high-level structures. In this chapter, we learned about the powerful in-memory AST representation that's used in Clang, including its C++ classes and APIs. This gives you a clear picture of the source code you are analyzing.

Furthermore, we learned and practiced a concise way to do pattern matching on AST – a crucial procedure for program analysis – via Clang's ASTMatcher. Familiarizing yourself with this technique can greatly improve your efficiency when it comes to filtering out interesting areas from the input source code. Last but not least, we learned how to write an AST plugin that makes it easier for you to integrate custom logic into the default Clang compilation pipeline.

In the next chapter, we will look at the **drivers** and **toolchains** in Clang. We will show you how they work and how to customize them.

8
Working with Compiler Flags and Toolchains

In the previous chapter, we learned how to process Clang's AST – one of the most common formats for analyzing programs. In addition, we learned how to develop an AST plugin, which is an easy way to insert custom logic into the Clang compilation pipeline. This knowledge will help you augment your skillset for tasks such as source code linting or finding potential security vulnerabilities.

In this chapter, we are ascending from specific subsystems and looking at the bigger picture – the compiler **driver** and **toolchain** that orchestrate, configure, and run individual LLVM and Clang components according to users' needs. More specifically, we will focus on how to add new compiler flags and how to create a custom toolchain. As we mentioned in *Chapter 5*, *Exploring Clang's Architecture*, compiler drivers and toolchains are often under-appreciated and have long been ignored. However, without these two important pieces of software, compilers will become extremely difficult to use. For example, users need to pass over *10* different compiler flags merely to build a simple *hello world* program, owing to the lack of flag translation. Users also need to run at least three different kinds of tools in order to create an executable to run, since there are no drivers or toolchains to help us invoke *assemblers* and *linkers*. In this chapter, you will learn how compiler drivers and toolchains work in Clang and how to customize them, which is extremely useful if you want to support Clang on a new operating system or architecture.

In this section, we will cover the following topics:

- Understanding drivers and toolchains in Clang

- Adding custom driver flags

- Adding a custom toolchain

Technical requirements

In this chapter, we are still relying on the `clang` executable, so make sure you build it, like this:

```
$ ninja clang
```

Since we are working with a driver, as we mentioned in *Chapter 5*, *Exploring Clang's Architecture*, you can use the -### command-line option to print out the frontend flags that have been translated from the driver, like so:

```
$ clang++ -### -std=c++11 -Wall hello_world.cpp -o hello_world
"/path/to/clang" "-cc1" "-triple" "x86_64-apple-macosx11.0.0"
"-Wdeprecated-objc-isa-usage" "-Werror=deprecated-objc-
isa-usage" "-Werror=implicit-function-declaration" "-emit-
obj" "-mrelax-all" "-disable-free" "-disable-llvm-verifier"
… "-fno-strict-return" "-masm-verbose" "-munwind-tables"
"-target-sdk-version=11.0" … "-resource-dir" "/Library/
Developer/CommandLineTools/usr/lib/clang/12.0.0" "-isysroot"
"/Library/Developer/CommandLineTools/SDKs/MacOSX.sdk" "-I/
usr/local/include" "-stdlib=libc++" … "-Wall" "-Wno-reorder-
init-list" "-Wno-implicit-int-float-conversion" "-Wno-c99-
```

```
designator" … "-std=c++11" "-fdeprecated-macro" "-fdebug-
compilation-dir" "/Users/Rem" "-ferror-limit" "19"
"-fmessage-length" "87" "-stack-protector" "1" "-fstack-
check" "-mdarwin-stkchk-strong-link" … "-fexceptions" …
"-fdiagnostics-show-option" "-fcolor-diagnostics" "-o" "/path/
to/temp/hello_world-dEadBeEf.o" "-x" "c++" "hello_world.cpp"…
```

Using this flag will *not* run the rest of the compilation but merely execute the driver and toolchain. This makes it a good way to verify and debug specific flags and check they are correctly propagated from the driver to the frontend.

Last but not least, in the last section of this chapter, *Adding a custom toolchain*, we will work on a project that can only run on Linux systems. Also, please install OpenSSL beforehand. It is usually available as a package in most Linux systems. For example, on Ubuntu, you can use the following command to install it:

```
$ sudo apt install openssl
```

We are only using the command-line utility, so there's no need to install any OpenSSL libraries that are normally used for development.

The code that will be used in this chapter can be found here: `https://github.com/PacktPublishing/LLVM-Techniques-Tips-and-Best-Practices-Clang-and-Middle-End-Libraries/tree/main/Chapter08`.

In the first section of this chapter, we will provide a brief introduction to Clang's driver and toolchain infrastructure.

Understanding drivers and toolchains in Clang

Before we talk about the compiler driver in Clang, it is necessary to highlight the fact that *compiling a piece of code* is never a *single* task (and not a simple one, either). In school, we were taught that a compiler consists of a **lexer**, a **parser**, sometimes came with an **optimizer**, and ended with an **assembly code printer**. While you still can see these stages in real-world compilers, they give you nothing but textual assembly code rather than an executable or library, as we would normally expect. Furthermore, this naïve compiler only provides limited flexibility – it can't be ported to any other operating systems or platforms.

To make this toy compiler more realistic and usable, many other *plumber* tools need to be put together, along with the core compiler: an **assembler** to transform assembly code into (binary format) object file, a **linker** to put multiple object files into an executable or library, and many other routines to resolve platform-specific configurations, such as data width, default header file paths, or **Application Binary Interfaces** (**ABIs**). Only with help from these *plumbers* can we use a compiler by just typing a few words:

```
$ clang hello_world.c -o hello_world
```

A **compiler driver** is software that organizes these *plumber* jobs. Despite having multiple different tasks to do during the compilation, we will only focus on two of the most important ones in this chapter – handling compiler flags and invoking the right tools on different platforms – which is what toolchains are designed for.

The following diagram shows the interactions between the driver, the toolchains, and the rest of the compiler:

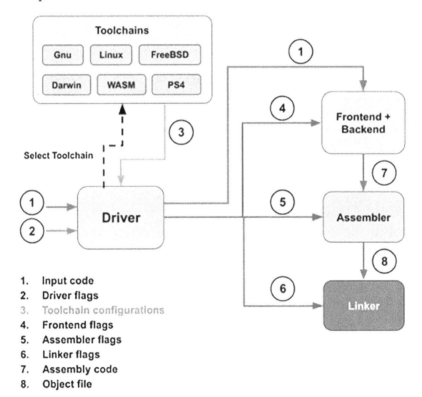

1. Input code
2. Driver flags
3. Toolchain configurations
4. Frontend flags
5. Assembler flags
6. Linker flags
7. Assembly code
8. Object file

Figure 8.1 – Typical workflow of Clang's driver, toolchains, and the rest of the compiler

As shown in the preceding diagram, Clang's driver acts as a *dispatcher* and distributes flags and workloads to each of the compilation phases, namely the frontend/backend, the assembler, and the linker. To give you a more concrete idea of what the flags for each of these phases look like, recall the `-###` compiler option we introduced at the beginning of this chapter. The (massive amount of) content that's printed by that option is the flags for the frontend (**4** in the preceding screenshot). For example, among those frontend flags, `-internal-isystem` carries the information about the system header path, including the path where the C/C++ standard library header files are stored. It is obvious that Clang's frontend needs to know where the standard library headers are stored, but as per your past experiences of using `clang` (or `gcc`), you rarely need to tell them where those headers are *explicitly* – the driver will do that for you. The same logic applies to the linking phase as well. Linkers usually need more than just an object file to properly generate an executable or a library. For example, they need to know where the C/C++ standard library's library files (`*.a` or `*.so` on Unix/Linux systems) are. In that case, Clang's driver will provide that information to the linkers via linker flags.

Flags and workloads – or *configurations*, in short – that are provided to individual compiler phases are *translated* from two sources: driver flags (**2** in the preceding diagram) and the selected toolchain (**3** in the preceding diagram). Driver flags are those provided by users via the command-line interface – that is, the *compiler flags*– such as `-c`, `-Wall`, and `-std=c++11`. In the next section, *Adding custom driver flags*, we will show you some examples of how Clang translates driver flags into frontend flags or even assembler/linker flags.

On the other hand, a **toolchain** is an entity that describes how input code should be compiled on **a specific platform**. Different hardware architectures and **operating systems (OS)** – *platforms*, for short – have their own way to build, load, and run programs. Take macOS X and Linux, for example. Although they both have a *Unix-like* environment, when building a program, the system (standard) libraries for macOS X always reside in Apple's XCode IDE package, whereas Linux usually stores them in normal folders such as `/usr/include` and `/usr/lib`. Also, macOS X uses an executable format called **Mach-O**, which is different from Linux's ELF format. This greatly affects how compilers (Clang) build the code.

For Clang to compile code for various platforms, it uses toolchains (which are effectively represented by the `ToolChain` C++ class internally) to encapsulate platform-specific information and configurations. In the early stage of compilation, Clang's driver selects a correct toolchain based on the system currently running (called the **host** system) or the users' preference – you can use the `-target=` driver flag to ask Clang to build a program for a specific platform that is different from the host system, which is effectively doing **cross-compiling**. Then, the driver will gather some platform-specific configurations from the selected toolchain before combining it with the aforementioned driver options and dispatching them to individual compiler phases via command-line flags. Note that different platforms usually use different assemblers and linkers. For example, macOS X can only use `ld64` and `lld` linkers for now, whereas Linux can use `ld` (BFD linker), `ld.gold`, and `lld` as linkers. Therefore, a toolchain should also specify what assembler and linker to use. In the last section of this chapter, *Adding a custom toolchain*, we will go through an example project to learn how Clang's toolchains work. Let's start our journey by learning how driver flags work in Clang.

Adding custom driver flags

In the previous section, we explained the role of the driver and toolchains in Clang. In this section, we are going to learn how Clang's driver does this translation by adding a custom driver flag to Clang. Again, we will go through the overview for this example project first before demonstrating the detailed steps in a separate section.

Project overview

The example project we will be using for this section is going to add a new driver flag so that when that flag is given by users, a header file will be *implicitly* included in the input code.

To be more specific, here, we have a header file – `simple_log.h` – shown in the following code that defines some simple APIs to print log messages:

```
#ifndef SIMPLE_LOG_H
#define SIMPLE_LOG_H
#include <iostream>
#include <string>
#ifdef SLG_ENABLE_DEBUG
inline void print_debug(const std::string &M) {
  std::cout << "[DEBUG] " << M << std::endl;
}
```

```
#endif


#ifdef SLG_ENABLE_ERROR
inline void print_error(const std::string &M) {
  std::cout << "[ERROR] " << M << std::endl;
}
#endif


#ifdef SLG_ENABLE_INFO
inline void print_info(const std::string &M) {
  std::cout << "[INFO] " << M << std::endl;
}
#endif


#endif
```

The goal here is to use these APIs in our code *without* writing the #include "simple_log.h" line to import the header file. And this feature will only be enabled when we give a custom driver flag, -fuse-simple-log, to clang. For example, let's write the following code, test.cc:

```
int main() {
  print_info("Hello world!!");
  return 0;
}
```

Despite its lack of any #include directives, it can still be compiled (with the -fuse-simple-log flag) and run without any problems:

```
$ clang++ -fuse-simple-log test.cc -o test
$ ./test
[INFO] Hello world!!
$
```

Moreover, we can use `-fuse-<log level>-simple-log` / `-fno-use-<log level>-simple-log` to include or exclude a function for a specific log level. For example, let's use the same preceding code snippets but add `-fno-use-info-simple-log` when we compile the code:

```
$ clang++ -fuse-simple-log -fno-use-info-simple-log test.cc -o test
test.cc:2:3: error: use of undeclared identifier 'print_info'
  print_info("Hello World!!");
  ^
1 error generated
$
```

The switch for each log printing function is simply controlled by its surrounding `#ifdef` statements in `simple_log.h`. For example, `print_info` will only be included if `SLG_ENABLE_INFO` is defined. Later, in the *Translating custom driver flags* section, we will show you how these macro definitions are toggled by driver flags.

Last but not least, you can specify a custom path to the `simple_log.h` file. By default, our feature will include `simple_log.h` in the current folder of the source code. You can change this by supplying `-fsimple-log-path=<file path>` or `-fuse-simple-log=<file path>`. For example, we want to use an alternative version of `simple_log.h` – `advanced_log.h`, which is stored in `/home/user` – which provides functions with the same interfaces but different implementations. Now, we can use the following commands:

```
$ clang++ -fuse-simple-log=/home/user/advanced_log.h test.cc -o test
[01/28/2021 20:51 PST] [INFO] Hello World!!
$
```

The following section will show you how to change the code in Clang's driver so that you can implement these features.

Declaring custom driver flags

First, we will lead you through the steps to *declare* custom driver flags such as `-fuse-simple-log` and `-fno-use-info-simple-log`. Then, we are going to *wire* those flags to the real frontend functionalities.

Clang uses **TableGen** syntax to declare all kinds of compiler flags – both driver flags and frontend flags.

> TableGen
>
> **TableGen** is a **Domain-Specific Language** (**DSL**) that's used for declaring structural and relational data. To learn more, please check out *Chapter 4, TableGen Development*.

All these flag declarations are put in `clang/include/clang/Driver/Options.td`. Take the common `-g` flag, for example, which tells you that you want to generate source-level debug information. For example, it has a declaration like this:

```
def g_Flag : Flag<["-"], "g">, Group<g_Group>,
  HelpText<"Generate source-level debug information">;
```

The TableGen record, `g_Flag`, is created from several TableGen classes: `Flag`, `Group`, and `HelpText`. Among them, we are most interested in `Flag`, whose template values (`["-"]` and `"g"`) describe the actual command-line flag format. Note that when we are declaring a *boolean* flag – the value of this flag is determined by its presence and no other values follow – as in this case, we inherit from the `Flag` class.

In cases where we want to declare a flag that has values that follow an equal sign ("="), we inherit from the `Joined` class. For example, the TableGen declaration for `-std=<C++ standard name>` looks like this:

```
def std_EQ : Joined<["-", "--"], "std=">, Flags<[CC1Option]>,
  ...;
```

Usually, the record names (`std_EQ`, in this case) for these kinds of flags have `_EQ` as their suffices.

Last but not least, the `Flags` (plural) class can be used to specify some properties. For example, `CC1Options` in the preceding snippet tells us that this flag can also be a frontend flag.

Now that we've learned how driver flags are generally declared, it is time to create our own:

1. First, we are going to deal with the `-fuse-simple-log` flag. Here is how we declare it:

    ```
    def fuse_simple_log : Flag<["-"], "fuse-simple-log">,
                          Group<f_Group>, Flags<[NoXarchOption]>;
    ```

    This snippet basically has no differences from the examples we used previously, except for the `Group` class and `NoXarchOption`. The former specifies the *logical* group this flag belongs to – for example, `f_Group` is for flags starting with `-f`. The latter tells us that this flag can *only* be used in the driver. You cannot, for example, pass it to the frontend (but how do we pass flags directly to the frontend? We will answer this question shortly, at the end of this section).

    Note that we only declare `-fuse-simple-log` here but not `-fuse-simple-log=<file path>` – that will be done in *another* flag that we will introduce shortly.

2. Next, we are dealing with `-fuse-<log level>-simple-log` and `-fno-use-<log level>-simple-log`. In both GCC and Clang, it is pretty common to see pairwise flags such as `-f<flag name>`/`-fno-<flag name>` to enable or disable a certain feature. Therefore, Clang provides a handy TableGen utility – `BooleanFFlag` – to make creating pairwise flags easier. Please see the declarations for `-fuse-error-simple-log`/`-fno-use-error-simple-log` in the following code:

    ```
    defm use_error_simple_log : BooleanFFlag<"use-error-
    simple-log">, Group<f_Group>, Flags<[NoXarchOption]>;
    ```

    `BooleanFFlag` is a *multiclass* (so make sure you use `defm` rather than `def` to create the TableGen record). Under the hood, it creates TableGen records for both `-f<flag name>` and `-fno-<flag name>` *at the same time*.

    Now that we've learned how `use_error_simple_log` was created, we can use the same trick to create TableGen records for other log levels:

    ```
    defm use_debug_simple_log : BooleanFFlag<"use-debug-
    simple-log">, Group<f_Group>, Flags<[NoXarchOption]>;
    defm use_info_simple_log : BooleanFFlag<"use-info-simple-
    log">, Group<f_Group>, Flags<[NoXarchOption]>;
    ```

3. Finally, we are declaring the `-fuse-simple-log=<file path>` and `-fsimple-log-path=<file path>` flags. In the previous steps, we were only dealing with boolean flags, but here, we are creating flags that have values that follow the equal sign, so we are using the `Joined` class we introduced earlier:

```
def fsimple_log_path_EQ : Joined<["-"], "fsimple-log-
path=">, Group<f_Group>, Flags<[NoXarchOption]>;
def fuse_simple_log_EQ : Joined<["-"], "fuse-simple-
log=">, Group<f_Group>, Flags<[NoXarchOption]>;
```

Again, flags with values will usually use `_EQ` in their TableGen record name suffix.

That wraps up all the necessary steps for declaring our custom driver flags. During Clang's building process, these TableGen directives will be translated into C++ enums and other utilities that are used by the driver. For example, `-fuse-simple-log=<file path>` will be represented by an enum; that is, `options::OPT_fuse_simple_log_EQ`. The next section will show you how to query these flags from all the command-line flags given by users and, most importantly, how to translate our custom flags into their frontend counterparts.

Translating custom driver flags

Recall that compiler drivers do a lot of things for users under the hood. For instance, they figure out the correct toolchain based on the compilation target and translate driver flags that have been designated by users, which is what we are going to do next. In our case here, we want to include the `simple_log.h` header file for users when our newly created `-fuse-simple-log` is given and define macro variables such as `SLG_ENABLE_ERROR` to include or exclude certain log printing functions, depending on the `-fuse-<log level>-simple-log`/`-fno-use-<log level>-simple-log` flags. More specifically, these tasks can be broken down into two parts:

- If `-fuse-simple-log` is given, we are translating it into a frontend flag:

```
-include "simple_log.h"
```

The `-include` frontend flag, as its name suggests, *implicitly* includes the designated file in the compiling source code.

Using the same logic, if `-fuse-simple-log=/other/file.h` or `-fuse-simple-log -fsimple-log-path=/other/file.h` are given, they will be translated into the following:

```
-include "/other/file.h"
```

- If either `-fuse-<log level>-simple-log` or `-fno-use-<log level>-simple-log` is given – for instance, `-fuse-error-simple-log` – it will be translated into the following:

  ```
  -D SLG_ENABLE_ERROR
  ```

  The `-D` flag implicitly defines a macro variable for the compiling source code.

  However, if only `-fuse-simple-only` is given, the flag will implicitly include all the log printing functions. In other words, `-fuse-simple-only` will not only be translated into the `-include` flag, as introduced in previous bullet point, but also the following flags:

  ```
  -D SLG_ENABLE_ERROR -D SLG_ENABLE_DEBUG -D SLG_ENABLE_
  INFO
  ```

  Let's say a combination of `-fuse-simple-log` and `-fno-use-<log level>-simple-log` are used together, for example:

  ```
  -fuse-simple-log -fno-use-error-simple-log
  ```

  They will be translated into the following code:

  ```
  -include "simple_log.h" -D SLG_ENABLE_DEBUG -D SLG_
  ENABLE_INFO
  ```

  Last but not least, we also allow the following combinations:

  ```
  -fuse-info-simple-log -fsimple-log-path="my_log.h"
  ```

  That is, we only enable a single log printing function without using `-fuse-simple-log` (instead of using the latter flag and subtracting two other log printing functions) and use a custom simple log header file. These driver flags will be translated into the following code:

  ```
  -include "my_log.h" -D SLG_ENABLE_INFO
  ```

  The aforementioned rules and combinations of flags can actually be handled in a pretty elegant way, albeit being complex at first glance. We will show you how to do this shortly.

Now that we have learned *what* frontend flags we are going to translate to, it is time to learn *how* to do these translations.

The place where many driver flags translations happen is inside the `driver::tools::Clang` C++ class. More specifically, this happens in its `Clang::ConstructJob` method, which is located in the `clang/lib/Driver/ToolChains/Clang.cpp` file.

> **About driver::tools::Clang**
>
> Some of the most prominent questions for this C++ class are probably, what *concept* does it represent? Why is it put under the folder named *ToolChains*? Does that mean it is also a toolchain? While we will answer these questions in detail in the next section, *Adding a custom toolchain*, for now, you can just think of it as the representative of Clang's frontend. This (kind of) explains why it is responsible for translating driver flags into frontend ones.

Here are the steps to translate our custom driver flags. The following code can be inserted anywhere within the `Clang::ConstructJob` method, before the `addDashXForInput` function is called, which starts to wrap up the translation process:

1. First, we are defining a help class – `SimpleLogOpts` – to carry our custom flag's information:

```
struct SimpleLogOpts {
  // If a certain log level is enabled
  bool Error = false,
       Info = false,
       Debug = false;
  static inline SimpleLogOpts All() {
    return {true, true, true};
  }
  // If any of the log level is enabled
  inline operator bool() const {
    return Error || Info || Debug;
  }
};
// The object we are going to work on later
SimpleLogOpts SLG;
```

The `bool` fields in `SimpleLogOpts` – `Error`, `Info`, and `Debug` – represent log levels that are enabled by our custom flags. We also define a helper function `SimpleLogOpts::All()` to create a `SimpleLogOpts` in which all log levels are enabled, and a `bool` type conversion operator such that we can use a cleaner syntax, shown here, to tell us if any of the levels are enabled:

```
if (SLG) {
  // At least one log level is enabled!
}
```

2. Let's handle the simplest case first – the `-fuse-simple-log` flag. In this step, we are only going to turn on all the log levels in `SLG` when we see a `-fuse-simple-log` flag.

    Inside the `Clang::ConstructJob` method, the driver flags given by users are stored in the `Args` variable (one of the arguments for `ConstructJob`), which is of the `ArgList` type. There are many ways to query `Args`, but here, since we only care about the *presence* of `-fuse-simple-log`, `hasArg` is the most suitable option:

    ```
    if (Args.hasArg(options::OPT_fuse_simple_log)) {
      SLG = SimpleLogOpts::All();
    }
    ```

    Each flag we declared in the previous code via TableGen syntax will be represented by a unique *enum* under the `options` namespace. In this case, the enum value is `OPT_fuse_simple_log`. The name of the enum value is usually `OPT_`, followed by the **TableGen record name** (that is, the name follows `def` or `defm`) when we were declaring the flag. The `ArgList::hasArg` function will return true if the given flag identifier is present in the input driver flags.

    In addition to `-fuse-simple-log`, we also need to turn on all the log levels when `-fuse-simple-log=<file path>` is given, even though we are only going to handle the file path that follows later. Thus, we will change the preceding snippet into the following:

    ```
    if (Args.hasArg(options::OPT_fuse_simple_log,
                    options::OPT_fuse_simple_log_EQ)) {
      SLG = SimpleLogOpts::All();
    }
    ```

    `ArgList::hasArg` can actually take multiple flag identifiers and return true if *any* of them are present in the input driver flags. And again, the `-fuse-simple-log=<...>` flag is represented by `OPT_fuse_simple_log_EQ` since its TableGen record name is `fuse_simple_log_EQ`.

3. Next, we are going to handle `-fuse-<log level>-simple-log`/`-fno-use-<log level>-simple-log`. Taking the error level, as an example (flags for other levels are used in the exact same way, so we are not showing them here), here, we are leveraging the `ArgList::hasFlag` function:

    ```
    SLG.Error = Args.hasFlag(options::OPT_fuse_error_simple_
    log, options::OPT_fno_use_error_simple_log, SLG.Error);
    ```

The hasFlag function will return true or false if the flag that's represented by the first (OPT_fuse_error_simple_log here) or second (OPT_fno_use_error_simple_log here) argument is present in the input driver flags, respectively.

If *neither* of the flags are present, hasFlag will return a default value that's designated by its third argument (SLG.Error, in this case).

Using this mechanism, we have already implemented some of the (complex) rule and flag combinations we mentioned earlier in this section:

a) The -fno-use-<log level>-simple-log flags can disable certain log printing function(s) when -fuse-simple-log – which effectively includes all the log printing functions in the first place – is present.

b) Even *without* the presence of -fuse-simple-log, we can still enable individual log printing functions by using the -fuse-<log level>-simple-log flag(s).

4. Currently, we are only playing around with the SimpleLogOpts data structure. Starting from the next step, we will start to generate frontend flags according to the SimpleLogOpts instance we have built so far. The first frontend flag we are generating here is -include <file path>. First, it only makes sense to proceed if at least one log level has been enabled. Therefore, we will wrap the generation of -include with an if statement by checking on SLG, as we explained earlier:

```
if (SLG) {
  CmdArgs.push_back("-include");
  …
}
```

The CmdArgs (a local variable – with a vector-like type – inside Clang::ConstructJob) is the place where we will put our **frontend** flags.

Note that you cannot push a frontend flag containing any *white space*. For instance, you cannot do something like this:

```
if (SLG) {
  CmdArgs.push_back("-include simple_log.h"); // Error
  …
}
```

This is because, eventually, this vector (`CmdArgs`) will be treated as `argv`, which we can see in the `main` function of C/C++, and any white space within a single argument will create failures when those arguments are realized.

Instead, we are pushing the path to a simple log header file *separately*, as follows:

```
if (SLG) {
  CmdArgs.push_back("-include");
  if (Arg *A = Args.getLastArg(options::OPT_fuse_simple_
  log_EQ, options::OPT_fsimple_log_path_EQ))
    CmdArgs.push_back(A->getValue());
  else
    CmdArgs.push_back("simple_log.h");
  ...
}
```

The `ArgList::getLastArg` function will retrieve the value (the last value, if there are multiple occurrences of the same flag), follow a given flag, and return null if none of those flags are present. For instance, in this case, the flag is `-fuse-simple-log=` (`-fsimple-log-path=` in the second argument is just the *alias* flag of the first one).

5. Finally, we are generating frontend flags that control which log printing functions should be enabled. Again, we are only showing the code for one of the log levels here since other levels are using the same approach:

```
if (SLG) {
  ...
  if (SLG.Error) {
    CmdArgs.push_back("-D");
    CmdArgs.push_back("SLG_ENABLE_ERROR");
  }
  ...
}
```

These are basically all the modifications that are required for our project. The final thing we must do before we move on is verify our work. Recall the `-###` command-line flag, which is used to print all the flags that are passed to the frontend. We are using it here to see if our custom driver flags are translated properly.

First, let's try this command:

```
$ clang++ -### -fuse-simple-log -c test.cc
```

The output should contain these strings:

```
"-include" "simple_log.h" "-D" "SLG_ENABLE_ERROR" "-D" "SLG_
ENABLE_INFO" "-D" "SLG_ENABLE_DEBUG"
```

Now, let's try the following command:

```
$ clang++ -### -fuse-simple-log=my_log.h -fno-use-error-simple-
log -c test.cc
```

Tthe output should contain these strings:

```
"-include" "my_log.h" "-D" "SLG_ENABLE_INFO" "-D" "SLG_ENABLE_
DEBUG"
```

Finally, let's use the following command:

```
$ clang++ -### -fuse-info-simple-log -fsimple-log-path=my_log.h
-c test.cc
```

The output should contain the following strings :

```
"-include" "my_log.h" "-D" "SLG_ENABLE_INFO"
```

In the last subsection of this section, we are going to talk about some miscellaneous ways to pass flags to the frontend.

Passing flags to the frontend

In the previous sections, we showed you the differences between driver flags and frontend flags, how they are related, and how Clang's driver translates the former into the latter. At this point, you might be wondering, can we skip through the driver and pass the flags directly to the frontend? What flags are we allowed to pass?

The short answer for the first question is *yes, and you have actually already done that several times in previous chapters*. Recall that in *Chapter 7, Handling AST*, we developed a plugin – more specifically, an AST plugin. We were using command-line arguments like the one shown here to load and run our plugin inside Clang:

```
$ clang++ -fplugin=MyPlugin.so \
          -Xclang -plugin -Xclang ternary-converter \
          -fsyntax-only test.cc
```

You might already find that, somehow, we need to precede a -Xclang flag before the -plugin and ternary-converter arguments. And the answer is simple: this is because -plugin (and its value, ternary-converter) is a *frontend-only* flag.

To pass a flag directly to the frontend, we can put -Xclang in front of it. But there is a caveat of using -Xclang: a single -Xclang will only relay *one* succeeding command-line argument (a string without any whitespace) to the frontend. In other words, you cannot rewrite the preceding plugin loading example like this:

```
# Error: `ternary-converter` will not be recognized
$ clang++ -fplugin=MyPlugin.so \
          -Xclang -plugin ternary-converter \
          -fsyntax-only test.cc
```

This is because -Xclang will only transfer -plugin to the frontend and leave ternary-converter behind, in which case Clang will fail to know which plugin to run.

Another way to pass flags directly to the frontend would be using -cc1. Recall that when we were using -### to print out the frontend flags that had been translated by the driver in the previous sections, among those frontend flags, the first one that followed the path to the clang executable was always -cc1. This flag effectively collects all the command-line arguments and sends them to the frontend. Though this looks handy – there's no need to prefix every flag we want to pass to the frontend with -Xclang anymore – be careful that you are not allowed to mix any *driver-only* flags inside that list of flags. For example, earlier in this section, when we were declaring our -fuse-simple-log flag in TableGen syntax, we annotated the flag with NoXarchOption, which stated that it can only be used by the driver. In that case, -fuse-simple-log cannot appear after -cc1.

This leads us to our final question: what flags can be used by either the driver or the frontend, and what flags are accepted by both? The answer can actually be seen via `NoXarchOption`, which was just mentioned. When declaring flags – either for the driver or the frontend – in TableGen syntax, you can use the `Flags<...>` TableGen class and its template parameters to enforce some constraints. For instance, using the following directives, you can *prevent* the `-foo` flag from being used by the driver:

```
def foo : Flag<["-"], "foo">, Flags<[NoDriverOption]>;
```

In addition to `NoXarchOption` and `NoDriverOption`, here are some other common annotations you can use in `Flags<...>`:

- `CoreOption`: States that this flag can be shared by both `clang` and `clang-cl`. `clang-cl` is an interesting driver that is compatible with the command-line interface (including command-line arguments) used by **MSVC** (the compiler framework used by Microsoft Visual Studio).

- `CC1Option`: States that this flag can be accepted by the frontend. It doesn't say it's a frontend-only flag, though.

- `Ignored`: States that this flag is going to be ignored by Clang's driver (but continue the compilation process). GCC has many flags that are not supported in Clang (either obsolete or just not applicable). However, Clang actually tries to *recognize* those flags but does nothing except show a warning message about a lack of implementation. The rationale behind this is we hope that Clang can be a *drop-in* replacement for GCC without the need to modify the existing building scripts in many projects (without this compatibility layer, Clang will terminate the compilation when it sees unknown flags).

In this section, we learned how to add custom flags for Clang's driver and implemented the logic to translate them into frontend flags. This skill is pretty useful when you want to toggle custom features in a more straightforward and clean way.

In the next section, we are going to learn the role of a toolchain and how it works in Clang by creating our own custom one.

Adding a custom toolchain

In the previous section, we learned how to add custom flags for the driver in Clang and learned how the driver translated them into flags that are accepted by the frontend. In this section, we are going to talk about the toolchain – an important module inside the driver that helps it adapt to different platforms.

Recall that in the first section of this chapter, *Understanding drivers and toolchains in Clang*, we showed the relationships between driver and toolchains in *Figure 8.1*: the driver chooses a proper toolchain based on the target platform before leveraging its knowledge to do the following:

1. Execute the correct *assembler*, *linker*, or any tool that is required for the target code's generation.

2. Pass *platform-specific* flags to the compiler, assembler, or linker.

This information is crucial for building the source code since each platform might have its own unique characteristics, such as system library paths and supported assembler/linker variants. Without them, a correct executable or library cannot even be generated.

This section hopes to teach you how to create Clang toolchains for custom platforms in the future. The toolchain framework in Clang is powerful enough to be adapted to a wide variety of use cases. For example, you can create a toolchain that resembles conventional compilers on Linux – including using GNU AS to assemble and GNU LD for linking – without you needing to make many customizations to a default library path or compiler flags. On the other hand, you can have an exotic toolchain that does not even use Clang to compile source code and uses a propriety assembler and linker with uncommon command-line flags. This section will try to use an example that catches the most common use cases without missing this framework's flexible aspect.

This section is organized as follows: as usual, we will start with an overview of the project we are going to work on. After that, we will break down our project workload into three parts – adding custom compiler options, setting up a custom assembler, and setting up a custom linker – before we put them together to wrap up this section.

> **System requirements**
>
> As another friendly reminder, the following project can only work on Linux systems. Please make sure OpenSSL is installed.

Project overview

We are going to create a toolchain called **Zipline**, which still uses Clang (its frontend and backend) to do normal compilation but encode the generated assembly code using **Base64** during the assembling phase, and package those Base64-encoded files into a **ZIP file** (or `.tarbell` file) during the linking phase.

> **Base64**
>
> **Base64** is an encoding scheme that is commonly used to convert binary into plain text. It can be easily transmitted in a context that does not support binary format (for example, HTTP headers). You can also apply Base64 to normal textual files, just like in our case.

This toolchain is basically useless in production environments. It's merely a demo that emulates common situations a developer might encounter when they're creating a new toolchain for custom platforms.

This toolchain is enabled by a custom driver flag, `-zipline/--zipline`. When the flag is provided, first, the compiler will implicitly add the `my_include` folder to your home directory as one of the header files searching the path. For example, recall that in the previous section, *Adding custom driver flags*, our custom `-fuse-simple-log` flag would implicitly include a header file, `simple_log.h`, in the input source code:

```
$ ls
main.cc simple_log.h
$ clang++ -fuse-simple-log -fsyntax-only main.cc
$ # OK
```

However, if `simple_log.h` is not in the current directory, as in the preceding snippet, we need to specify its full path via another flag:

```
$ ls .
# No simple_log.h in current folder
main.cc
$ clang++ -fuse-simple-log=/path/to/simple_log.h -fsyntax-only
main.cc
$ # OK
```

With the help of Zipline, you can put `simple_log.h` inside `/home/<user name>/my_include`, and the compiler will find it:

```
$ ls .
# No simple_log.h in current folder
main.cc
$ ls ~/my_include
simple_log.h
$ clang++ -zipline -fuse-simple-log -fsyntax-only main.cc
$ # OK
```

The second feature of Zipline is that the `clang` executable will compile the source code into assembly code that's encoded by Base64 under the `-c` flag, which *was* supposed to assemble the assembly file – coming out from the compiler – into an object file. Here is an example command:

```
$ clang -zipline -c test.c
$ file test.o
test.o: ASCII text # Not (binary) object file anymore
$ cat test.o
CS50ZXh0CgkuZmlsZQkidGVzdC5jYyIKCS
5nbG9ibAlfWjNmb29pCgkucDJhbGln
bgk0LCAweDkwCgkudHlwZQlfWjNmb29p
LEBmdW5jdGlvbgpfWjNmb29pOgoJLmNm
... # Base64 encoded contents
$
```

The preceding `file` command showed that the generated file, `test.o`, from the previous invocation of `clang`, is no longer a binary format object file. The content of this file is now a Base64-encoded version of the assembly code that was generated from the compiler's backend.

Finally, Zipline replaces the original linking stage with a custom one that packages and compresses the aforementioned Base64-encoded assembly files into a `.zip` file. Here is an example:

```
$ clang -zipline test.c -o test.zip
$ file test.zip
test.zip: Zip archive, at least v2.0 to extract
$
```

If you unzip `test.zip`, you will find that those extracted files are Base64-encoded assembly files, as we mentioned earlier.

Alternatively, we can use Linux's `tar` and `gzip` utilities to package and compress them in Zipline. Let's look at an example:

```
$ clang -zipline -fuse-ld=tar test.c -o test.tar.gz
$ file test.tar.gz
test.tar.gz: gzip compressed data, from Unix, original size…
$
```

By using the existing `-fuse-ld=<linker name>` flag, we can choose between using `zip` or `tar` and `gzip` for our custom linking phase.

In the next section, we are going to create the skeleton code for this toolchain and show you how to add an additional folder to the header file searching path.

Creating the toolchain and adding a custom include path

In this section, we are going to create the skeleton for our Zipline toolchain and show you how to add an extra include folder path – more specifically, an extra **system include path** – to the compilation stage within Zipline. Here are the detailed steps:

1. Before we add a real toolchain implementation, don't forget that we are going to use a custom driver flag, `-zipline/--zipline`, to enable our toolchain. Let's use the same skill we learned in the previous section, *Adding custom driver flags*, to do that. Inside `clang/include/clang/Driver/Options.td`, we will add the following lines:

   ```
   // zipline toolchain
   def zipline : Flag<["-", "--"], "zipline">,
                 Flags<[NoXarchOption]>;
   ```

   Again, `Flag` tells us this is a boolean flag and `NoXarchOption` tells us that this flag is driver-only. We will use this driver flag shortly.

2. Toolchains in Clang are represented by the `clang::driver::ToolChain` class. Each toolchain supported by Clang is derived from it, and their source files are put under the `clang/lib/Driver/ToolChains` folder. We are going to create two new files there: `Zipline.h` and `Zipline.cpp`.

3. For `Zipline.h`, let's add the following skeleton code first:

   ```
   namespace clang {
   namespace driver {
   namespace toolchains {
   struct LLVM_LIBRARY_VISIBILITY ZiplineToolChain
     : public Generic_ELF {
     ZiplineToolChain(const Driver &D, const llvm::Triple
       &Triple, const llvm::opt::ArgList &Args)
       : Generic_ELF(D, Triple, Args) {}
   ```

```
  ~ZiplineToolChain() override {}

  // Disable the integrated assembler
  bool IsIntegratedAssemblerDefault() const override
    { return false; }
  bool useIntegratedAs() const override { return false; }

  void
  AddClangSystemIncludeArgs(const llvm::opt::ArgList
    &DriverArgs, llvm::opt::ArgStringList &CC1Args)
    const override;

protected:
  Tool *buildAssembler() const override;
  Tool *buildLinker() const override;
};
} // end namespace toolchains
} // end namespace driver
} // end namespace clang
```

The class we created here, ZiplineToolChain, is derived from Generic_ELF, which is a subclass of ToolChain that's specialized for systems that use ELF as its execution format – including Linux. In addition to the parent class, there are three important methods that we are going to implement in this or later sections: AddClangSystemIncludeArgs, buildAssembler, and buildLinker.

4. The buildAssembler and buildLinker methods generate Tool instances that represent the **commands** or programs to be run in the assembling and linking stages, respectively. We will cover them in the following sections. Now, we are going to implement the AddClangSystemIncludeArgs method. Inside Zipline. cpp, we will add its method body:

```
  void ZiplineToolChain::AddClangSystemIncludeArgs(
                        const ArgList &DriverArgs,
                        ArgStringList &CC1Args) const {
    using namespace llvm;
    SmallString<16> CustomIncludePath;
    sys::fs::expand_tilde("~/my_include",
                        CustomIncludePath);
```

```
    addSystemInclude(DriverArgs,
                    CC1Args, CustomIncludePath.c_str());
}
```

The only thing we are doing here is calling the addSystemInclude function with the full path to the my_include folder located in the home directory. Since each user's home directory is different, we are using the sys::fs::expand_tilde helper function to expand ~/my_include – where ~ represents the home directory in Linux and Unix systems – in the absolute path. The addSystemInclude function, on the other hand, helps you add "-internal-isystem" "/path/to/my_include" flags to the list of all the frontend flags. The -internal-isystem flag is used for designating folders of system header files, including standard library headers and some platform-specific header files.

5. Last but not least, we need to teach the driver to use the Zipline toolchain when it sees our newly created -zipline/--zipline driver flag. We are going to modify the Driver::getToolChain method inside clang/lib/Driver/Driver.cpp to do so. The Driver::getToolChain method contains a huge switch case for selecting different toolchains based on the target operating system and hardware architecture. Please navigate to the code handling the Linux system; we are going to add an extra branch condition there:

```
const ToolChain
&Driver::getToolChain(const ArgList &Args,
                    const llvm::Triple &Target) const {
  ...
  switch (Target.getOS()) {
  case llvm::Triple::Linux:
  ...
    else if (Args.hasArg(options::OPT_zipline))
      TC = std::make_unique<toolchains::ZiplineToolChain>
      (*this, Target, Args);
  ...
    break;
  case ...
  case ...
  }
}
```

The extra `else-if` statement basically says that if the target OS is Linux, then we will use Zipline when the `-zipline/--zipline` flag is given.

With that, you have added the skeleton of Zipline and successfully told the driver to use Zipline when a custom driver flag is given. On top of that, you've also learned how to add extra system library folders to the header file search path.

In the next section, we are going to create a custom assembling stage and connect it to the toolchain we created here.

Creating a custom assembling stage

As we mentioned in the *Project overview* section, instead of doing regular assembling to convert assembly code into an object file in the assembling stage of Zipline, we are invoking a program to convert the assembly file we generated from Clang into its Base64-encoded counterpart. Before we go deeper into its implementation, let's learn how each of these *stages* in a toolchain is represented.

In the previous section, we learned that a toolchain in Clang is represented by the `ToolChain` class. Each of these `ToolChain` instances is responsible for telling the driver what *tool* to run in each compilation stage – namely compiling, assembling, and linking. And this information is encapsulated inside a `clang::driver::Tool` type object. Recall the `buildAssembler` and `buildLinker` methods in the previous section; they return the very `Tool` type objects that depict the actions to perform and the tool to run in the assembling and linking stages, respectively. In this section, we will show you how to implement the `Tool` object that's returned by `buildAssembler`. Let's get started:

1. Let's go back to `Zipline.h` first. Here, we are adding an extra class, `Assembler`, inside the `clang::driver::tools::zipline` namespace:

```
namespace clang {
namespace driver {
namespace tools {

namespace zipline {
struct LLVM_LIBRARY_VISIBILITY Assembler : public Tool {
  Assembler(const ToolChain &TC)
    : Tool("zipline::toBase64", "toBase64", TC) {}

  bool hasIntegratedCPP() const override { return false;
}
```

```
    void ConstructJob(Compilation &C, const JobAction &JA,
                      const InputInfo &Output,
                      const InputInfoList &Inputs,
                      const llvm::opt::ArgList &TCArgs,
                      const char *LinkingOutput) const
                      override;
  };
  } // end namespace zipline
  } // end namespace tools


  namespace toolchains {
  struct LLVM_LIBRARY_VISIBILITY ZiplineToolChain … {
    …
  };
  } // end namespace toolchains
  } // end namespace driver
  } // end namespace clang
```

Be careful because the newly created `Assembler` resides in the
`clang::driver::tools::zipline` namespace, while `ZiplineToolChain`,
which we created in the previous section, is in `clang::driver::toolchains`.

The `Assembler::ConstructJob` method is where we will put our logic for
invoking Base64 encoding tools.

2. Inside `Zipline.cpp`, we will implement the method body of
    `Assembler::ConstructJob`:

```
    void
    tools::zipline::Assembler::ConstructJob(Compilation &C,
                            const JobAction &JA,
                            const InputInfo &Output,
                            const InputInfoList &Inputs,
                            const ArgList &Args,
                            const char *LinkingOutput)
                            const {
                            ArgStringList CmdArgs;
                            const InputInfo &II =
                            Inputs[0];
```

```
std::string Exec =
  Args.MakeArgString(getToolChain().
  GetProgramPath("openssl"));


// opeenssl base64 arguments
CmdArgs.push_back("base64");
CmdArgs.push_back("-in");
CmdArgs.push_back(II.getFilename());
CmdArgs.push_back("-out");
CmdArgs.push_back(Output.getFilename());


C.addCommand(
  std::make_unique<Command>(
        JA, *this, ResponseFileSupport::None(),
        Args.MakeArgString(Exec), CmdArgs,
        Inputs, Output));
}
```

We are using OpenSSL to do the Base64 encoding, and the command we hope to run is as follows:

```
$ openssl base64 -in <input file> -out <output file>
```

The job of the `ConstructJob` method is building a *program invocation* to run the previous command. It is realized by the `C.addCommand(...)` function call at the very end of `ConstructJob`. The `Command` instance that's passed to the `addCommand` call represents the concrete command to be run during the assembling stage. It contains necessary information such as the path to the program executable (the `Exec` variable) and its arguments (the `CmdArgs` variable).

For the `Exec` variable, the toolchain has provided a handy utility, the `GetProgramPath` function, to resolve the absolute path of an executable for you.

The way we build arguments for `openssl` (the `CmdArgs` variable), on the other hand, is very similar to the thing we did in the *Adding custom driver flags* section: translating driver flags (the `Args` argument) and the input/output file information (the `Output` and `Inputs` argument) into a new set of command-line arguments and storing them in `CmdArgs`.

3. Finally, we connect this `Assembler` class with `ZiplineToolChain` by implementing the `ZiplineToolChain::buildAssembler` method:

```
Tool *ZiplineToolChain::buildAssembler() const {
    return new tools::zipline::Assembler(*this);
}
```

These are all the steps we need to follow to create a `Tool` instance that represents the command to run during the linking stage of our Zipline toolchain.

Creating a custom linking stage

Now that we've finished the assembler stage, it's time to move on to the next stage – the linking stage. We are going to use the same approach we used in the previous section; that is, we will create a custom `Tool` class representing the linker. Here are the steps:

1. Inside `Zipline.h`, create a `Linker` class that is derived from `Tool`:

```
namespace zipline {
struct LLVM_LIBRARY_VISIBILITY Assembler : public Tool {
...
};

struct LLVM_LIBRARY_VISIBILITY Linker : public Tool {
  Linker(const ToolChain &TC)
     : Tool("zipeline::zipper", "zipper", TC) {}

  bool hasIntegratedCPP() const override { return false;
}
  bool isLinkJob() const override { return true; }

  void ConstructJob(Compilation &C, const JobAction &JA,
                    const InputInfo &Output,
                    const InputInfoList &Inputs,
                    const llvm::opt::ArgList &TCArgs,
                    const char *LinkingOutput) const
                    override;

private:
  void buildZipArgs(const JobAction&, const InputInfo&,
```

```
                          const InputInfoList&,
                          const llvm::opt::ArgList&,
                          llvm::opt::ArgStringList&) const;

    void buildTarArgs(const JobAction&,
                          const InputInfo&,
                          const InputInfoList&,
                          const llvm::opt::ArgList&,
                          llvm::opt::ArgStringList&) const;
  };
  } // end namespace zipline
```

In this Linker class, we also need to implement the ConstructJob method
to tell the driver what to execute during the linking stage. Differently from
Assembler, since we need to support both the zip and tar + gzip packaging/
compression schemes, we will add two extra methods, buildZipArgs and
buildTarArgs, to handle argument building for each.

2. Inside Zipline.cpp, we'll focus on the implementation of
   Linker::ConstructJob first:

```
void
tools::zipline::Linker::ConstructJob(Compilation &C,
                          const JobAction &JA,
                          const InputInfo &Output,
                          const InputInfoList &Inputs,
                          const ArgList &Args,
                          const char *LinkingOutput) const
{
  ArgStringList CmdArgs;
  std::string Compressor = "zip";
  if (Arg *A = Args.getLastArg(options::OPT_fuse_ld_EQ))
    Compressor = A->getValue();
  std::string Exec = Args.MakeArgString(
      getToolChain().GetProgramPath(Compressor.c_str()));

  if (Compressor == "zip")
    buildZipArgs(JA, Output, Inputs, Args, CmdArgs);
  if (Compressor == "tar" || Compressor == "gzip")
```

```
        buildTarArgs(JA, Output, Inputs, Args, CmdArgs);
    else
        llvm_unreachable("Unsupported compressor name");

    C.addCommand(
        std::make_unique<Command>(
            JA, *this, ResponseFileSupport::None(),
            Args.MakeArgString(Exec),
            CmdArgs, Inputs, Output));
}
```

In this custom linking stage, we hope to use either the `zip` command or the `tar` command – depending on the `-fuse-ld` flag specified by users – to package all the (Base64-encoded) files generated by our custom `Assembler`.

The detailed command format for both `zip` and `tar` will be explained shortly. From the preceding snippet, we can see that the thing we are doing here is similar to `Assembler::ConstructJob`. The `Exec` variable carries the absolute path to either the `zip` or `tar` program; the `CmdArgs` variable, which is populated by either `buildZipArgs` or `buildTarArgs`, which will be explained later, carries the command-line arguments for the tool (`zip` or `tar`).

The biggest difference compared to `Assembler::ConstructJob` is that the command to execute can be designated by the `-fuse-ld` flag that's supplied by users. Thus, we are using the skill we learned about in the *Adding custom driver flags* section to read that driver flag and set up the command.

3. If your users decide to package files in a ZIP file (which is the default scheme, or you can specify it explicitly via `-fuse-ld=zip`), we are going to run the following command:

    ```
    $ zip <output zip file> <input file 1> <input file 2>…
    ```

    Therefore, we will build our `Linker::buildZipArgs` method, which constructs an argument list for the preceding command, as follows:

    ```
    void
    tools::zipline::Linker::buildZipArgs(const JobAction &JA,
                                         const InputInfo &Output,
                                         const InputInfoList &Inputs,
                                         const ArgList &Args,
                                         ArgStringList &CmdArgs)
    ```

```
                                      const {
  // output file
  CmdArgs.push_back(Output.getFilename());
  // input files
  AddLinkerInputs(getToolChain(), Inputs, Args, CmdArgs,
JA);
}
```

The CmdArgs argument of Linker::buildZipArgs will be where we'll export our results. While we are still using the same way to fetch the output filename (via Output.getFilename()), since a linker might accept multiple inputs at a time, we are leveraging another helper function, AddLinkerInputs, to add all the input filenames to CmdArgs for us.

4. If your users decide to use the tar + gzip packaging scheme (via the -fuse-ld=tar or -fuse-ld=gzip flags), we are going to run the following command:

    ```
    $ tar -czf <output tar.gz file> <input file 1> <input
    file 2>…
    ```

    Therefore, we will build our Linker::buildTarArgs method, which constructs an argument list for the previous command, as follows:

    ```
    void
    tools::zipline::Linker::buildTarArgs(const JobAction &JA,
                            const InputInfo &Output,
                            const InputInfoList &Inputs,
                            const ArgList &Args,
                            ArgStringList &CmdArgs)
                            const {
      // arguments and output file
      CmdArgs.push_back("-czf");
      CmdArgs.push_back(Output.getFilename());
      // input files
      AddLinkerInputs(getToolChain(), Inputs, Args, CmdArgs,
        JA);
    }
    ```

Just like `buildZipArgs`, we grab the output filename via `Output.getFilename()` and add all the input filenames, using `AddLinkerInput`, into `CmdArgs`.

5. Last but not least, let's connect our `Linker` to `ZiplineToolChain`:

```
Tool *ZiplineToolChain::buildLinker() const {
  return new tools::zipline::Linker(*this);
}
```

That's all of the steps for implementing a custom linking phase for our Zipline toolchain.

Now that we have created the necessary components for the Zipline toolchain, we can execute our custom features – encode the source files and package them into an archive – when users select this toolchain. In the next section, we are going to learn how to verify these functionalities.

Verifying the custom toolchain

To test the functionalities we implemented in this chapter, we can run the example commands depicted in the project overview or we can leverage the `-###` driver flag again to dump all the expected compiler, assembler, and linker command details.

So far, we've learned that the `-###` flag will show all the frontend flags that have been translated by the driver. But actually, it will also show the assembler and linker commands that have been scheduled to run. For instance, let's invoke the following command:

```
$ clang -### -zipline -c test.c
```

Since the `-c` flag always tries to run the assembler over the assembly file generated by Clang, our custom assembler (that is, the Base64 encoder) within Zipline will be triggered. Therefore, you will see an output similar to the following:

```
$ clang -### -zipline -c test.c
"/path/to/clang" "-cc1" …
"/usr/bin/openssl" "base64" "-in" "/tmp/test_ae4f5b.s" "-out"
"test.o"
$
```

The line starting with `/path/to/clang -cc1` contains the frontend flags we learned about earlier. The line that follows is the assembler invocation command. This, in our case, runs `openssl` to perform Base64 encoding.

Note that the weird `/tmp/test_ae4f5b.s` filename is the temporary file that's created by the driver to accommodate the assembly code that's generated by the compiler.

Using the same trick, we can verify our custom linker stage, as follows:

```
$ clang -### -zipline test.c -o test.zip
"/path/to/clang" "-cc1" …
"/usr/bin/openssl" "base64" "-in" "/tmp/test_ae4f5b.s" "-out"
"/tmp/test_ae4f5b.o"
"/usr/bin/zip" "test.zip" "/tmp/test_ae4f5b.o"
$
```

Since the `-o` flag was used in the previous command, Clang will build a complete executable from `test.c` involving the assembler and the linker. Therefore, our custom linking stage is up here due to the `zip` command taking the result (the `/tmp/test_ae4f5b.o` file) from the previous assembling stage. Feel free to add the `-fuse-ld=tar` flag to see the `zip` command replace the `tar` command with a completely different argument list.

In this section, we showed you how to create a toolchain for Clang's driver. This is a crucial skill for supporting Clang on custom or new platforms. We also learned that the toolchain framework in Clang is flexible and can handle a variety of tasks that are required by the target platform.

Summary

In this chapter, we started by introducing Clang's driver and the role of the toolchain – the module that provides platform-specific information such as the supported assemblers and linkers – that assisted it. Then, we showed you one of the most common ways to customize the driver – adding a new driver flag. After that, we talked about the toolchain and, most importantly, how to create a custom one. These skills are really useful when you want to create a new feature in Clang (or even LLVM) and need a custom compiler flag to enable it. Also, the ability to develop a custom toolchain is crucial for supporting Clang on new operating systems, or even new hardware architecture.

This is the final chapter of the second part of this book. Starting from the next chapter, we will talk about LLVM's middle end – the *platform-independent* program analysis and optimization framework.

Exercises

1. It is common to override the assembling and linking stage since different platforms tend to support different assemblers and linkers. However, is it possible to override the *compiling* stage (which is Clang)? If it is possible, how can we do this? Why may people wish to do this?

2. When we were working on `tools::zipline::Linker::ConstructJob`, we simply use `llvm_unreachable` to bail out the compilation process if a user provides an unsupported compressor name through the `-fuse-ld` flag. Can we replace it with Clang's **diagnostic** framework, which we learned about in *Chapter 7, Handling AST*, to print out better messages?

3. Just like we can use `-Xclang` to pass flags directly to the frontend, we can also pass assembler-specific or linker-specific flags directly to the assembler or linker using driver flags such as `-Wa` (for the assembler) or `-Wl` (for the linker). How can we consume those flags in our custom assembler and linker stages within Zipline?

Section 3: "Middle-End" Development

Target-independent transformations and analyses, also known as the "middle-end" in LLVM, is the core of the entire framework. Though you can find most resources online for this part, there are still so many hidden features that can improve your development experience, and so many pitfalls that will tip you over the edge out of nowhere. In this section, we will prepare you for the journey of LLVM middle-end development. This section includes the following chapters:

- *Chapter 9, Working with PassManager and AnalysisManager*
- *Chapter 10, Processing LLVM IR*
- *Chapter 11, Gearing Up with Support Utilities*
- *Chapter 12, Learning LLVM IR Instrumentation*

9
Working with PassManager and AnalysisManager

In the previous section of this book, *Frontend Development*, we began with an introduction to the internals of Clang, which is LLVM's official frontend for the C family of programming languages. We went through various projects, involving skills and knowledge, that can help you to deal with problems that are tightly coupled with source code.

In this part of the book, we will be working with **LLVM IR** – a target-independent **intermediate representation** (**IR**) for compiler optimization and code generation. Compared to Clang's **Abstract Syntax Tree** (**AST**), LLVM IR provides a different level of abstraction by encapsulating extra execution details to enable more powerful program analyses and transformations. In addition to the design of LLVM IR, there is a mature ecosystem around this IR format, which provides countless resources, such as libraries, tools, and algorithm implementations. We will cover a variety of topics in LLVM IR, including the most common LLVM Pass development, using and writing program analysis, and the best practices and tips for working with LLVM IR APIs. Additionally, we will review more advanced skills such as **Program Guided Optimization** (**PGO**) and sanitizer development.

In this chapter, we are going to talk about writing a transformation **Pass** and program analysis for the new **PassManager**. LLVM Pass is one of the most fundamental, and crucial, concepts within the entire project. It allows developers to encapsulate program processing logic into a modular unit that can be freely composed with other Passes by the **PassManager**, depending on the situation. In terms of the design of the Pass infrastructure, LLVM has actually gone through an overhaul of both PassManager and AnalysisManager to improve their runtime performance and optimization quality. The new PassManager uses a quite different interface for its enclosing Passes. This new interface, however, is not backward-compatible to the legacy one, meaning you cannot run legacy Passes in the new PassManager and vice versa. What is worse, there aren't many learning resources online that talk about this new interface, even though, now, they are enabled, by default, in both LLVM and Clang. The content of this chapter will fill this gap and provide you with an up-to-date guide to this crucial subsystem in LLVM.

In this chapter, we will cover the following topics:

- Writing an LLVM Pass for the new PassManager

- Working with the new AnalysisManager

- Learning instrumentations in the new PassManager

With the knowledge learned from this chapter, you should be able to write an LLVM Pass, using the new Pass infrastructure, to transform or even optimize your input code. You can also further improve the quality of your Pass by leveraging the analysis data provided by LLVM's program analysis framework.

Technical requirements

In this chapter, we will, primarily, use a command-line utility, called `opt`, to test our Passes. You can build it using a command like this:

```
$ ninja opt
```

The code example for this chapter can be found at `https://github.com/PacktPublishing/LLVM-Techniques-Tips-and-Best-Practices-Clang-and-Middle-End-Libraries/tree/main/Chapter09`.

Writing an LLVM Pass for the new PassManager

A **Pass in LLVM** is the basic unit that is required to perform certain actions against LLVM IR. It is similar to a single production step in a factory, where the products that need to be processed are LLVM IR and the factory workers are the Passes. In the same way that a normal factory usually has multiple manufacturing steps, LLVM also consists of multiple Passes that are executed in sequential order, called the **Pass pipeline**. *Figure 9.1* shows an example of the Pass pipeline:

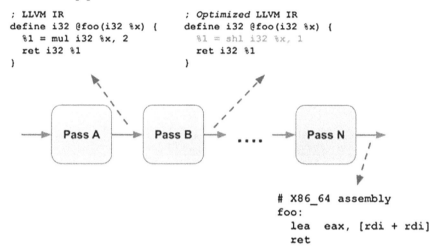

```
; LLVM IR                          ; Optimized LLVM IR
define i32 @foo(i32 %x) {          define i32 @foo(i32 %x) {
  %1 = mul i32 %x, 2                 %1 = shl i32 %x, 1
  ret i32 %1                         ret i32 %1
}                                  }
```

```
# X86_64 assembly
foo:
    lea   eax, [rdi + rdi]
    ret
```

Figure 9.1 – An example of the LLVM Pass pipeline and its intermediate results

In the preceding diagram, multiple Passes are arranged in a straight line. The LLVM IR for the `foo` function is processed by one Pass after another. **Pass B**, for instance, performs code optimization on `foo` and replaces an arithmetic multiplication (`mul`) by 2 with left shifting (`shl`) by 1, which is considered easier than multiplication in most hardware architectures. In addition, this figure also illustrates that the **code generation** steps are modeled as Passes. Code generation in LLVM transforms LLVM IR, which is target independent, into assembly code for certain hardware architecture (for example, **x86_64** in *Figure 9.1*). Each detailed procedure, such as the register allocation, instruction selection, or instruction scheduling, is encapsulated into a single Pass and is executed in a certain order.

Code generation Passes

Passes for code generation have a different API than normal LLVM IR Passes. Additionally, during the code generation phase, LLVM IR is actually converted into another kind of IR, called **Machine IR** (**MIR**). However, in this chapter, we will only be covering LLVM IR and its Passes.

This Pass pipeline is conceptually managed by an infrastructure called **PassManager**. PassManager owns the plan – their execution order, for example – to run these Passes. Conventionally, we actually use the terms *Pass pipeline* and *PassManager* interchangeably since they have nearly identical missions. In the *Learning instrumentations in the new PassManager* section, we will go into more detail about the pipeline itself and discuss how to customize the execution order of these enclosing Passes.

Code transformations in modern compilers can be complex. Because of this, multiple transformation Passes might need the same set of program information, which is called **analysis** in LLVM, in order to do their work. Furthermore, to achieve maximum efficiency, LLVM also *caches* this analysis data so that it can be reused if possible. However, since a transformation Pass might change the IR, some cached analysis data, which was previously collected, might be outdated after running that Pass. To solve these challenges, in addition to PassManager, LLVM has also created **AnalysisManager** to manage everything related to program analysis. We will go deeper into AnalysisManager in the *Working with the new AnalysisManager* section.

As mentioned in the introduction of this chapter, LLVM has gone through a series of overhauls on its Pass and PassManager (and AnalysisManager) infrastructure. The new infrastructure runs faster and generates results with better quality. Nevertheless, the new Pass differs in many places from the old one; we will briefly explain these differences along the way. However, aside from that, we will only be discussing the new Pass infrastructure, by default, for the rest of the chapter.

In this section, we will show you how to develop a simple Pass for the new PassManager. As usual, we will begin with a description of the sample project we are about to use. Then, we will show you the steps to create a Pass that can be dynamically loaded from a plugin into the Pass pipeline, which was mentioned earlier, using the `opt` utility.

Project overview

In this section, the sample project we are using is called **StrictOpt**. It is a Pass and Pass plugin that adds a `noalias` attribute to every function parameter that has a pointer type. Effectively, it adds a `restrict` keyword to the function parameters in C code. First, let's explain what the `restrict` keyword does.

The restrict keyword in C and C++

The `restrict` keyword was introduced in C99. However, it doesn't have a counterpart in C++. Nevertheless, mainstream compilers such as Clang, GCC, and MSVS all support the same functionality in C++. For example, in Clang and GCC, you can use `__restrict__` or `__restrict` in C++ code and it has the same effect as `restrict` in C.

The `restrict` keyword can also be used alongside pointer type variables in C. In the most common cases, it is used with pointer type function parameters. The following is an example:

```
int foo(int* restrict x, int* restrict y) {
    *x = *y + 1;
    return *y;
}
```

Essentially, this additional attribute tells the compiler that argument x will never point to the *same memory region* as argument y. In other words, programmers can use this keyword to *persuade* the compilers that they will *never* call the `foo` function, as follows:

```
...
// Programmers will NEVER write the following code
int main() {
    int V = 1;
    return foo(&V, &V);
}
```

The rationale behind this is that if the compiler knows that two pointers – in this case, the two pointer arguments – will never point to the same memory region, it can do more *aggressive* optimizations. To give you a more concrete understanding of this, if you compare the assembly code of the `foo` function with and without the `restrict` keyword, the latter version takes five instructions to execute (on x86_64):

```
foo:
    mov    eax, dword ptr [rsi]
    add    eax, 1
    mov    dword ptr [rdi], eax
    mov    eax, dword ptr [rsi]
    ret
```

The version with the `restrict` keyword added only takes four instructions:

```
foo:
    mov    eax, dword ptr [rsi]
    lea    ecx, [rax + 1]
    mov    dword ptr [rdi], ecx
    ret
```

Although the difference here seems subtle, in the version without `restrict`, the compiler needs to insert an extra memory load to assure that the last argument `*y` (in the original C code) always reads the latest value. This extra cost might gradually accumulate in a more complex code base and, eventually, create a performance bottleneck.

Now, you have learned how `restrict` works and its importance for ensuring good performance. In LLVM IR, there is also a corresponding directive to model the `restrict` keyword: the `noalias` attribute. This attribute is attached to the pointer function parameters if hints such as `restrict` have been given by programmers in the original source code. For example, the `foo` function (with the `restrict` keywords) can be translated into the following LLVM IR:

```
define i32 @foo(i32* noalias %0, i32* noalias %1) {
  %3 = load i32, i32* %1
  %4 = add i32 %3, 1
  store i32 %4, i32* %0
  ret i32 %3
}
```

Furthermore, we can also generate the LLVM IR code of the `foo` function *without* `restrict` in C code, as follows:

```
define i32 @foo(i32* %0, i32* %1) {
  %3 = load i32, i32* %1
  %4 = add i32 %3, 1
  store i32 %4, i32* %0
  %5 = load i32, i32* %1
  ret i32 %5
}
```

Here, you will find that there is an extra memory load (as shown in the highlighted instruction of the preceding snippet), which is similar to what happened to the assembly examples from earlier. That is, LLVM is unable to perform more aggressive optimization to remove that memory load since it's not sure whether those pointers overlap each other.

In this section, we are going to write a Pass to add a `noalias` attribute to every pointer argument of a function. The Pass will be built as a plugin, and once it's loaded into `opt`, users can use the `--passes` argument to explicitly trigger `StrictOpt`, as follows:

```
$ opt --load-pass-plugin=StrictOpt.so \
      --passes="function(strict-opt)" \
      -S -o - test.ll
```

Alternatively, we can make `StrictOpt` run before other optimizations if the optimization level is greater or equal to `-O3`. The following is an example:

```
$ opt -O3 --enable-new-pm \
      --load-pass-plugin=StrictOpt.so \
      -S -o - test.ll
```

We will show you how to switch between these two modes shortly.

A demo-only Pass

Note that `StrictOpt` is merely a demo-only Pass, and adding `noalias` to every pointer function argument is absolutely *not* the thing you should do in real-world use cases. This is because it might break the **correctness** of the target program.

In the next section, we will show you detailed steps of how to create this Pass.

Writing the StrictOpt Pass

The following instructions will take you through the process of developing the core Pass logic before covering how to register `StrictOpt` into the Pass pipeline dynamically:

1. We only have two source files this time: `StrictOpt.h` and `StrictOpt.cpp`. In the former file, we place the skeleton of the `StrictOpt` Pass:

    ```
    #include "llvm/IR/PassManager.h"
    struct StrictOpt : public PassInfoMixin<StrictOpt> {
      PreservedAnalyses run(Function &F,
                            FunctionAnalysisManager &FAM);
    };
    ```

The Pass we are writing here is a *function Pass*; namely, it runs on the Function IR unit. The run method is the primary entry point for this Pass, which we are going to fill in later. It takes two arguments: a Function class that we will work on and a FunctionAnalysisManager class that can give you analysis data. It returns a PreservedAnalyses instance, which tells PassManager (and AnalysisManager) what analysis data was *invalidated* by this Pass.

If you have prior experience in writing LLVM Pass for the *legacy* PassManager, you might find several differences between the legacy Pass and the new Pass:

a) The Pass class no longer derives from one of the FunctionPass, ModulePass, or LoopPass. Instead, the Passes running on different IR units are all deriving from PassInfoMixin<YOUR_PASS>. In fact, deriving from PassInfoMixin is *not* even a requirement for a functional Pass anymore – we will leave this as an exercise for you.

b) Instead of *overriding* methods, such as runOnFunction or runOnModule, you will define a normal class member method, run (be aware that run does *not* have an override keyword that follows), which operates on the desired IR unit.

Overall, the new Pass has a cleaner interface compared to the legacy one. This difference also allows the new PassManager to have less overhead runtime.

2. To implement the skeleton from the previous step, we are heading to StrictOpt.cpp. In this file, first, we create the following method definition:

```
#include "StrictOpt.h"
using namespace llvm;
PreservedAnalyses StrictOpt::run(Function &F,
                          FunctionAnalysisManager &FAM) {
  return PreservedAnalyses::all(); // Just a placeholder
}
```

The returned PreservedAnalyses::all() instance is just a placeholder that will be removed later.

3. Now, we are finally creating the code to add a noalias attribute to the pointer function arguments. The logic is simple: for each Argument instance in a Function class, attach noalias if it fulfills the criteria:

```
// Inside StrictOpt::run…
bool Modified = false;
for (auto &Arg : F.args()) {
```

```
    if (Arg.getType()->isPointerTy() &&
        !Arg.hasAttribute(Attribute::NoAlias)) {
      Arg.addAttr(Attribute::NoAlias);
      Modified |= true;
    }
  }
```

The args() method of the Function class will return a range of Argument instances representing all of the formal parameters. We check each of their types to make sure there isn't an existing noalias attribute (which is represented by the Attribute::NoAlias enum). If everything looks good, we use addAttr to attach noalias.

Here, the Modified flag here records whether any of the arguments were modified in this function. We will use this flag shortly.

4. Certain analysis data might become outdated after a transformation Pass since the latter might change the program's IR. Therefore, when writing a Pass, we need to return a PreservedAnalyses instance to show which analysis was affected and should be subject to recalculation. While there are many analyses available in LLVM, we don't need to enumerate each of them. Instead, there are some handy utility functions to create PreservedAnalyses instances, representing *all analyses* or *none of the analyses*, such that we only need to subtract or add (un) affected analysis from it. Here is what we do in StrictOpt:

```
    #include "llvm/Analysis/AliasAnalysis.h"
    …
    // Inside StrictOpt::run…
    auto PA = PreservedAnalyses::all();
    if (Modified)
      PA.abandon<AAManager>();
    return PA;
```

Here, we first create a PreservedAnalyses instance, PA, which represents *all analyses*. Then, if the Function class we are working on here has been modified, we *discard* the AAManager analysis via the abandon method. AAManager represents the **alias analysis** in LLVM.

Without going into the details of this, the alias analysis asks whether two pointers point to the same memory region, or whether the memory regions they are pointing to overlap with each other. The `noalias` attribute we are discussing here has strong relations with this analysis since they're working on a nearly identical problem. Therefore, if any new `noalias` attribute was generated, all the cached alias analysis data would be outdated. This is why we invalidate it using `abandon`.

Note that you can always return a `PreservedAnalyses::none()` instance, which tells AnalysisManager to mark *every* analysis as outdated if you are not sure what analyses have been affected. This comes at a cost, of course, since AnalysisManager then needs to spend extra effort to recalculate the analyses that might contain expensive computations.

5. The core logic of `StrictOpt` is essentially finished. Now, we are going to show you how to dynamically register the Pass into the pipeline. In `StrictOpt.cpp`, we create a special global function, called `llvmGetPassPluginInfo`, with an outline like this:

```
extern "C" ::llvm::PassPluginLibraryInfo LLVM_ATTRIBUTE_
WEAK
llvmGetPassPluginInfo() {
  return {
    LLVM_PLUGIN_API_VERSION, "StrictOpt", "v0.1",
    [] (PassBuilder &PB) {…}
  };
}
```

This function returns a `PassPluginLibraryInfo` instance, which contains various pieces of information such as the plugin API version (`LLVM_PLUGIN_API_VERSION`) and the Pass name (`StrictOpt`). One of its most important fields is a lambda function that takes a single `PassBuilder&` argument. In that particular function, we are going to insert our `StrictOpt` into a proper position within the Pass pipeline.

`PassBuilder`, as its name suggests, is an entity LLVM that is used to build the Pass pipeline. In addition to its primary job, which involves configuring the pipeline according to the optimization level, it also allows developers to insert Passes into some of the places in the pipeline. Furthermore, to increase its flexibility, `PassBuilder` allows you to specify a *textual* description of the pipeline you want to run by using the `--passes` argument on `opt`, as we have seen previously. For instance, the following command will run `InstCombine`, `PromoteMemToReg`, and `SROA` (**SROA: Scalar Replacement of Aggregates**) in sequential order:

```
$ opt --passes="instcombine,mem2reg,sroa" test.ll -S -o -
```

What we are going to do in this step is ensure that after the plugin has been loaded, `opt` will run our Pass if `strict-opt` appears in the `--passes` argument, as follows:

```
$ opt --passes="strict-opt" test.ll -S -o -
```

To do this, we leverage the `registerPipelineParsingCallback` method in `PassBuilder`:

```
...
[](PassBuilder &PB) {
  using PipelineElement = typename
PassBuilder::PipelineElement;
  PB.registerPipelineParsingCallback(
    [](StringRef Name,
       FunctionPassManager &FPM,
ArrayRef<PipelineElement>){
      if (Name == "strict-opt") {
        FPM.addPass(StrictOpt());
        return true;
      }
      return false;
    });
}
```

The registerPipelineParsingCallback method takes another lambda callback as the argument. This callback is invoked whenever PassBuilder encounters an unrecognized Pass name while parsing the textual pipeline representation. Therefore, in our implementation, we simply insert our StrictOpt pass into the pipeline via FunctionPassManager::addPass when the unrecognized Pass name, that is, the Name parameter, is strict-opt.

6. Alternatively, we also want to trigger our StrictOpt at the beginning of the Pass pipeline without using the textual pipeline description, as we described in the *Project overview* section. This means that the Pass will be run before other Passes after it is loaded into opt using the following command:

```
$ opt -O2 --enable-new-pm \
    --load-pass-plugin=StrictOpt.so test.ll -S -o -
```

(The --enable-new-pm flag in the preceding command forced opt to use the new PassManager since it's still using the legacy one by default. We haven't used this flag before because --passes implicitly enables the new PassManager under the hood.)

To do this, instead of using
PassBuilder::registerPipelineParsingCallback
to register a custom (pipeline) parser callback, we are going to use
registerPipelineStartEPCallback to handle this. Here is the alternative version of the code snippet from the previous step:

```
...
[] (PassBuilder &PB) {
  using OptimizationLevel
    = typename PassBuilder::OptimizationLevel;
  PB.registerPipelineStartEPCallback(
    [] (ModulePassManager &MPM, OptimizationLevel OL) {
      if (OL.getSpeedupLevel() >= 2) {
        MPM.addPass(
          createModuleToFunctionPassAdaptor(StrictOpt()));
      }
    });
}
```

There are several things worth noting in the preceding snippet:

- The `registerPipelineStartEPCallback` method we are using here registers a callback that can customize certain places in the Pass pipeline, called **extension points** (**EPs**). The EP we are going to customize here is one of the earliest points in the pipeline.

- In comparison to the lambda callback we saw in `registerPipelineParsingCallback`, the lambda callback for `registerPipelineStartEPCallback` only provides `ModulePassManager`, rather than `FunctionPassManager`, to insert our `StrictOpt` Pass, which is a function Pass. We are using `ModuleToFunctionPassAdapter` to overcome this issue.

  `ModuleToFunctionPassAdapter` is a module Pass that can run a given function Pass over a module's enclosing functions. It is suitable for running a function Pass in contexts where only `ModulePassManager` is available, such as in this scenario. The `createModuleToFunctionPassAdaptor` function highlighted in the preceding code is used to create a new `ModuleToFunctionPassAdapter` instance from a specific function Pass.

- Finally, in this version, we are only enabling `StrictOpt` when the optimization level is greater or equal to `-O2`. Therefore, we leverage the `OptimizationLevel` argument passing into the lambda callback to determine whether we want to insert `StrictOpt` into the pipeline or not.

  With these Pass registration steps, we have also learned how to trigger our `StrictOpt` without explicitly specifying the textual Pass pipeline.

To summarize, in this section, we learned the essentials of the LLVM Pass and Pass pipeline. Through the `StrictOpt` project, we have learned how to develop a Pass – which is also encapsulated as a plugin – for the new PassManager and how to dynamically register it against the Pass pipeline in `opt` in two different ways: first, by triggering the Pass explicitly via a textual description, and second, by running it at a certain time point (EP) in the pipeline. We also learned how to invalidate analyses depending on the changes made in the Pass. These skills can help you develop high-quality and modern LLVM Passes to process IR in a composable fashion with maximum flexibility. In the next section, we will dive into the program analysis infrastructure of LLVM. This greatly improves the capability of normal LLVM transformation Passes.

Working with the new AnalysisManager

Modern compiler optimizations can be complex. They usually require lots of information from the target program in order to make correct decisions and optimal transformations. For example, in the *Writing an LLVM Pass for the new PassManager* section, LLVM used the `noalias` attribute to calculate memory aliasing information, which might eventually be used to remove redundant memory loads.

Some of this information – called **analysis**, in LLVM – is expensive to evaluate. In addition, a single analysis might also depend on other analyses. Therefore, LLVM creates an **AnalysisManager** component to handle all tasks related to program analysis in LLVM. In this section, we are going to show you how to use AnalysisManager in your own Passes for the sake of writing more powerful and sophisticated program transformations or analyses. We will also use a sample project, **HaltAnalyzer**, to drive our tutorial here. The next section will provide you with an overview of HaltAnalyzer before moving on to the detailed development steps.

Overview of the project

HaltAnalyzer is set up in a scenario where target programs are using a special function, `my_halt`, that terminates the program execution when it is called. The `my_halt` function is similar to the `std::terminate` function, or the `assert` function when its sanity check fails.

The job of HaltAnalyzer is to analyze the program to find basic blocks that are *guaranteed to be unreachable* because of the `my_halt` function. To be more specific, let's take the following C code as an example:

```c
int foo(int x, int y) {
  if (x < 43) {
    my_halt();
    if (y > 45)
      return x + 1;
    else {
      bar();
      return x;
    }
  } else {
    return y;
  }
}
```

Because my_halt was called at the beginning of the true block for the if (x < 43) statement, the code highlighted in the preceding snippet will never be executed (that is, my_halt stopped all of the program executions before even getting to those lines).

HaltAnalyzer should identify these basic blocks and print out warning messages to stderr. Just like the sample project from the previous section, HaltAnalyzer is also a function Pass wrapped inside a plugin. Therefore, if we use the preceding snippet as the input to our HaltAnalyzer Pass, it should print out the following messages:

```
$ opt --enable-new-pm --load-pass-plugin ./HaltAnalyzer.so \
        --disable-output ./test.ll
[WARNING] Unreachable BB: label %if.else
[WARNING] Unreachable BB: label %if.then2
$
```

The %if.else and %if.then2 strings are just names for the basic blocks in the if (y > 45) statement (you might see different names on your side). Another thing worth noting is the --disable-output command-line flag. By default, the opt utility will print out the binary form of LLVM IR (that is, the LLVM bitcode) anyway unless users redirect the output to other places via the -o flag. Using the aforementioned flag is merely to tell opt not to do that since we are not interested in the final content of LLVM IR (because we are not going to modify it) this time.

Although the algorithm of HaltAnalyzer seems pretty simple, writing it from scratch might be a pain. That's why we are leveraging one of the analyses provided by LLVM: the **Dominator Tree (DT)**. The concept of **Control Flow Graph** (CFG) domination has been taught in most entry-level compiler classes, so we are not going to explain it in depth here. Simply speaking, if we say a basic block *dominates* another block, every execution flow that arrives at the latter is guaranteed to go through the former first. A DT is one of the most important and commonly used analyses in LLVM; most control flow-related transformations cannot live without it.

Putting this idea into HaltAnalyzer, we are simply looking for all of the basic blocks that are dominated by the basic blocks that contain a function call to my_halt (we are excluding the basic blocks that contain the my_halt call sites from the warning messages). In the next section, we will show you detailed instructions on how to write HaltAnalyzer.

Writing the HaltAnalyzer Pass

In this project, we will only create a single source file, `HaltAnalyzer.cpp`. Most of the infrastructure, including `CMakeListst.txt`, can be reused from the `StrictOpt` project in the previous section:

1. Inside `HaltAnalyzer.cpp`, first, we create the following Pass skeleton:

```
class HaltAnalyzer : public PassInfoMixin<HaltAnalyzer> {
    static constexpr const char* HaltFuncName = "my_halt";
    // All the call sites to "my_halt"
    SmallVector<Instruction*, 2> Calls;
    void findHaltCalls(Function &F);

public:
    PreservedAnalyses run(Function &F,
                        FunctionAnalysisManager &FAM);
};
```

   In addition to the `run` method that we saw in the previous section, we are creating an additional method, `findHaltCalls`, which will collect all of the `Instruction` calls to `my_halt` in the current function and store them inside the `Calls` vector.

2. Let's implement `findHaltCalls` first:

```
void HaltAnalyzer::findHaltCalls(Function &F) {
  Calls.clear();
  for (auto &I : instructions(F)) {
    if (auto *CI = dyn_cast<CallInst>(&I)) {
      if (CI->getCalledFunction()->getName() ==
          HaltFuncName)
        Calls.push_back(&I);
    }
  }
}
```

   This method uses `llvm::instructions` to iterate through every `Instruction` call in the current function and check them one by one. If the `Instruction` call is a `CallInst` – representing a typical function call site – and the callee name is `my_halt`, we will push it into the `Calls` vector for later use.

Function name mangling

Be aware that when a line of C++ code is compiled into LLVM IR or native code, the name of any symbol – including the function name – will be different from what you saw in the original source code. For example, a simple function that has the name of *foo* and takes no argument might have *_Z3foov* as its name in LLVM IR. We call such a transformation in C++ **name mangling**. Different platforms also adopt different name mangling schemes. For example, in Visual Studio, the same function name becomes *?foo@@YAHH@Z* in LLVM IR.

---

3. Now, let's go back to the `HaltAnalyzer::run` method. There are two things we are going to do. We will collect the call sites to `my_halt` via `findHaltCalls`, which we just wrote, and then retrieve the DT analysis data:

```
#include "llvm/IR/Dominators.h"
...
PreservedAnalyses
HaltAnalyzer::run(Function &F, FunctionAnalysisManager
&FAM) {
    findHaltCalls(F);
    DominatorTree &DT = FAM.
      getResult<DominatorTreeAnalysis>(F);
    ...
}
```

The highlighted line in the preceding snippet is the main character of this section. It shows us how to leverage the provided `FunctionAnalysisManager` type argument to retrieve specific analysis data (in this case, `DominatorTree`) for a specific `Function` class.

Although, so far, we have (kind of) used the words *analysis* and *analysis data* interchangeably, in a real LLVM implementation, they are actually two different entities. Take the DT that we are using here as an example:

a) **DominatorTreeAnalysis** is a C++ class that evaluates dominating relationships from the given `Function`. In other words, it is the one that *performs* the analysis.

b) **DominatorTree** is a C++ class that represents the *result* generated from `DominatorTreeAnalysis`. This is just static data that will be cached by AnalysisManager until it is invalidated.

Furthermore, LLVM asks every analysis to clarify its affiliated result type via the `Result` member type. For example, `DominatorTreeAnalysis::Result` is equal to `DominatorTree`.

To make this even more formal, to associate the analysis data of an analysis class, `T`, with a `Function` variable, `F`, we can use the following snippet:

```
// `FAM` is a FunctionAnalysisManager
typename T::Result &Data = FAM.getResult<T>(F);
```

4. After we retrieve `DominatorTree`, it's time to find all of the basic blocks dominated by the `Instruction` call sites that we collected earlier:

```
PreservedAnalyses
HaltAnalyzer::run(Function &F, FunctionAnalysisManager
&FAM) {
  ...
  SmallVector<BasicBlock*, 4> DomBBs;
  for (auto *I : Calls) {
    auto *BB = I->getParent();
    DomBBs.clear();
    DT.getDescendants(BB, DomBBs);

    for (auto *DomBB : DomBBs) {
    // excluding the block containing `my_halt` call site
      if (DomBB != BB) {
        DomBB->printAsOperand(
              errs() << "[WARNING] Unreachable BB: ");
        errs() << "\n";
      }
    }
  }
  return PreservedAnalyses::all();
}
```

By using the `DominatorTree::getDescendants` method, we can retrieve all of the basic blocks dominated by a `my_halt` call site. Note that the results from `getDescendants` will also contain the block you put into the query (in this case, the block containing the `my_halt` call sites), so we need to exclude it before printing the basic block name using the `BasicBlock::printAsOperand` method.

With the ending of the returning `PreservedAnalyses::all()`, which tells AnalysisManager that this Pass does not invalidate any analysis since we don't modify the IR at all, we will wrap up the `HaltAnalyzer::run` method here.

5. Finally, we need to dynamically insert our HaltAnalyzer Pass into the Pass pipeline. We are using the same method that we did in the last section, by implementing the `llvmGetPassPluginInfo` function and using `PassBuilder` to put our Pass at one of the EPs in the pipeline:

```
extern "C" ::llvm::PassPluginLibraryInfo LLVM_ATTRIBUTE_
WEAK
llvmGetPassPluginInfo() {
  return {
    LLVM_PLUGIN_API_VERSION, "HaltAnalyzer", "v0.1",
    [](PassBuilder &PB) {
      using OptimizationLevel
        = typename PassBuilder::OptimizationLevel;
      PB.registerOptimizerLastEPCallback(
        [](ModulePassManager &MPM, OptimizationLevel OL)
{

          MPM.addPass(createModuleToFunctionPassAdaptor
            (HaltAnalyzer()));
        });
    }
  };
}
```

In comparison to `StrictOpt` in the previous section, we are using `registerOptimizerLastEPCallback` to insert HaltAnalyzer *after* all of the other optimization Passes. The rationale behind this is that some optimizations might move basic blocks around, so prompting warnings too early might not be very useful. Nevertheless, we are still leveraging `ModuletoFunctionPassAdaptor` to wrap around our Pass; this is because `registerOptimizerLastEPCallback` only provides `ModulePassManager` for us to add our Pass, which is a function Pass.

These are all the necessary steps to implement our HaltAnalyzer. Now you have learned how to use LLVM's program analysis infrastructure to obtain more information about the target program in an LLVM Pass. These skills can provide you with more insight into IR when you are developing a Pass. In addition, this infrastructure allows you to reuse high-quality, off-the-shelf program analysis algorithms from LLVM instead of recreating the wheels by yourself. To browse all of the available analyses provided by LLVM, the `llvm/include/llvm/Analysis` folder in the source tree is a good starting point. Most of the header files within this folder are standalone analysis data files that you can use.

In the final section of this chapter, we will show you some diagnosis techniques that are useful for debugging an LLVM Pass.

Learning instrumentations in the new PassManager

PassManager and AnalysisManager in LLVM are complicated pieces of software. They manage interactions between hundreds of Passes and analyses, and it can be a challenge when we try to diagnose a problem caused by them. In addition, it's really common for a compiler engineer to fix crashes in the compiler or **miscompilation** bugs. In those scenarios, useful instrumentation tools that provide insights to Passes and the Pass pipeline can greatly improve the productivity of fixing those problems. Fortunately, LLVM has already provided many of those tools.

Miscompilation

Miscompilation bugs usually refer to logical issues in the **compiled program**, which were introduced by compilers. For example, an overly aggressive compiler optimization removes certain loops that shouldn't be removed, causing the compiled software to malfunction, or mistakenly reorder memory barriers and create *race conditions* in the generated code.

We will introduce a single tool at a time in each of the following sections. Here is the list of them:

- Printing Pass pipeline details
- Printing changes to the IR after each Pass
- Bisecting the Pass pipeline

These tools can interact purely in the command-line interface of opt. In fact, you can also create *your own* instrumentation tools (without even changing the LLVM source tree!); we will leave this as an exercise for you.

Printing Pass pipeline details

There are many different **optimization levels** in LLVM, that is, the -O1, -O2, or -Oz flags we are familiar with when using clang (or opt). Each optimization level is running a *different set of Passes* and arranging them in *different orders*. In some cases, this might greatly affect the generated code, in terms of performance or correctness. Therefore, sometimes, it's crucial to know these configurations in order to gain a clear understanding of the problems we are going to deal with.

To print out all the Passes and the order they are currently running in inside opt, we can use the --debug-pass-manager flag. For instance, given the following C code, test.c, we will see the following:

```
int bar(int x) {
   int y = x;
   return y * 4;
}
int foo(int z) {
   return z + z * 2;
}
```

We first generate the IR for it using the following command:

```
$ clang -O0 -Xclang -disable-O0-optnone -emit-llvm -S test.c
```

> **The -disable-O0-optnone flag**
>
> By default, `clang` will attach a special attribute, `optnone`, to each function under the `-O0` optimization level. This attribute will prevent any further optimization on the attached functions. Here, the `-disable-O0-optnone` (frontend) flag is preventing `clang` from attaching to this attribute.

Then, we use the following command to print out all of the Passes running under the optimization level of `-O2`:

```
$ opt -O2 --disable-output --debug-pass-manager test.ll
Starting llvm::Module pass manager run.
…
Running pass: Annotation2MetadataPass on ./test.ll
Running pass: ForceFunctionAttrsPass on ./test.ll
…
Starting llvm::Function pass manager run.
Running pass: SimplifyCFGPass on bar
Running pass: SROA on bar
Running analysis: DominatorTreeAnalysis on bar
Running pass: EarlyCSEPass on bar
…
Finished llvm::Function pass manager run.
…
Starting llvm::Function pass manager run.
Running pass: SimplifyCFGPass on foo
…
Finished llvm::Function pass manager run.
Invalidating analysis: VerifierAnalysis on ./test.ll
…
$
```

The preceding command-line output tells us that `opt` first runs a set of *module-level* optimizations; the ordering of those Passes (for example, `Annotation2MetadataPass` and `ForceFunctionAttrsPass`) are also listed. After that, a sequence of *function-level* optimizations is performed on the `bar` function (for example, `SROA`) before running those optimizations on the `foo` function. Furthermore, it also shows the analyses used in the pipeline (for example, `DominatorTreeAnalysis`), as well as prompting us with a message regarding they became invalidated (by a certain Pass).

To sum up, `--debug-pass-manager` is a useful tool to peek into the Passes and their ordering run by the Pass pipeline at a certain optimization level. Knowing this information can give you a big picture of how Passes and analyses interact with the input IR.

Printing changes to the IR after each Pass

To understand the effects of a particular transformation Pass on your target program, one of the most straightforward ways is to compare the IR before and after it is processed by that Pass. To be more specific, in most cases, we are interested in the *changes* made by a particular transformation Pass. For instance, if LLVM mistakenly removes a loop that it shouldn't do, we want to know *what* Pass did that, and *when* the removal happened in the Pass pipeline.

By using the `--print-changed` flag (and some other supported flags that we will introduce shortly) with `opt`, we can print out the IR after each Pass if it was ever modified by that Pass. Using the `test.c` (and its IR file, `test.ll`) example code from the previous paragraph, we can use the following command to print changes, if there are any, made by each Pass:

```
$ opt -O2 --disable-output --print-changed ./test.ll
*** IR Dump At Start: ***
...
define dso_local i32 @bar(i32 %x) #0 {
entry:
  %x.addr = alloca i32, align 4
  %y = alloca i32, align 4
  ...
  %1 = load i32, i32* %y, align 4
  %mul = mul nsw i32 %1, 4
  ret i32 %mul
}
```

```
...
*** IR Dump After VerifierPass (module) omitted because no
change ***
...
...
*** IR Dump After SROA *** (function: bar)
; Function Attrs: noinline nounwind uwtable
define dso_local i32 @bar(i32 %x) #0 {
entry:
  %mul = mul nsw i32 %x, 4
  ret i32 %mul
}
...
$
```

Here, we have only shown a small amount of output. However, in the highlighted part of the snippet, we can see that this tool will first print out the original IR (IR Dump At Start), then show the IR after it is processed by each Pass. For example, the preceding snippet shows that the bar function has become much shorter after the SROA Pass. If a Pass didn't modify the IR at all, it will omit the IR dump to reduce the amount of noise.

Sometimes, we are only interested in the changes that have happened on a *particular set of functions*, say, the foo function, in this case. Instead of printing the *change log* of the entire module, we can add the --filter-print-funcs=<function names> flag to only print IR changes for a subset of functions. For example, to only print IR changes for the foo function, you can use the following command:

```
$ opt -O2 --disable-output \
        --print-changed --filter-print-funcs=foo ./test.ll
```

Just like --filter-print-funcs, sometimes, we only want to see changes made by a *particular set of Passes*, say, the SROA and InstCombine Passes. In that case, we can add the --filter-passes=<Pass names> flag. For example, to view only the content that is relevant to SROA and InstCombine, we can use the following command:

```
$ opt -O2 --disable-output \
        --print-changed \
        --filter-passes=SROA,InstCombinePass ./test.ll
```

Now you have learned how to print the IR differences among all the Passes in the pipeline, with additional filters that can further focus on a specific function or Pass. In other words, this tool can help you to easily observe the *progression* of changes throughout the Pass pipeline and quickly spot any traces that you might be interested in. In the next section, we will learn how to debug problems raised in the code optimization by *bisecting* the Pass pipeline.

Bisecting the Pass pipeline

In the previous section, we introduced the `--print-changed` flag, which prints out the *IR change log* throughout the Pass pipeline. We also mentioned that it is useful to call out changes that we are interested in; for instance, an invalid code transformation that caused miscompilation bugs. Alternatively, we can also **bisect** the Pass pipeline to achieve a similar goal. To be more specific, the `--opt-bisect-limit=<N>` flag in opt bisects the Pass pipeline by *disabling* all Passes except the first N ones. The following command shows an example of this:

```
$ opt -O2 --opt-bisect-limit=5 -S -o - test.ll
BISECT: running pass (1) Annotation2MetadataPass on module (./
test.ll)
BISECT: running pass (2) ForceFunctionAttrsPass on module (./
test.ll)
BISECT: running pass (3) InferFunctionAttrsPass on module (./
test.ll)
BISECT: running pass (4) SimplifyCFGPass on function (bar)
BISECT: running pass (5) SROA on function (bar)
BISECT: NOT running pass (6) EarlyCSEPass on function (bar)
BISECT: NOT running pass (7) LowerExpectIntrinsicPass on
function (bar)
BISECT: NOT running pass (8) SimplifyCFGPass on function (foo)
BISECT: NOT running pass (9) SROA on function (foo)
BISECT: NOT running pass (10) EarlyCSEPass on function (foo)
...
define dso_local i32 @bar(i32 %x) #0 {
entry:
  %mul = mul nsw i32 %x, 4
  ret i32 %mul
}
define dso_local i32 @foo(i32 %y) #0 {
```

```
entry:
  %y.addr = alloca i32, align 4
  store i32 %y, i32* %y.addr, align 4
  %0 = load i32, i32* %y.addr, align 4
  %1 = load i32, i32* %y.addr, align 4
  %mul = mul nsw i32 %1, 2
  %add = add nsw i32 %0, %mul
  ret i32 %add
}
$
```

(Note that this is different from examples shown in the previous sections; the preceding command has printed both messages from `--opt-bisect-limit` and the final textual IR.)

Since we implemented the `--opt-bisect-limit=5` flag, the Pass pipeline only ran the first five Passes. As you can see from the diagnostic messages, SROA was applied on `bar` but not the `foo` function, leaving the final IR of `foo` less optimal.

By changing the number that follows `--opt-bisect-limit`, we can adjust the cut point until certain code changes appear or a certain bug is triggered (for example, a crash). This is particularly useful as an *early filtering step* to narrow down the original problem to a smaller range of Passes in the pipeline. Furthermore, since it uses a numeric value as the parameter, this feature fits perfectly to automating environments such as automatic crash reporting tools or performance regression tracking tools.

In this section, we introduced several useful instrumentation tools in `opt` for debugging and diagnosing the Pass pipeline. These tools can greatly improve your productivity when it comes to fixing problems, such as compiler crashes, performance regressions (on the target program), and miscompilation bugs.

Summary

In this chapter, we learned how to write an LLVM Pass for the new PassManager and how to use program analysis data within a Pass via the AnalysisManager. We also learned how to leverage various instrumentation tools to improve the development experiences while working with the Pass pipeline. With the skills gained from this chapter, you can now write a Pass to process LLVM IR, which can be used to transform or even optimize a program.

These topics are some of the most fundamental and crucial skills to learn before starting on any IR level transformation or analysis task. If you have been working with the legacy PassManager, these skills can also help you to migrate your code to the new PassManager system, which has now been enabled by default.

In the next chapter, we will show you various tips along with the best practices that you should know when using the APIs of LLVM IR.

Questions

1. In the `StrictOpt` example of the *Writing an LLVM Pass for the new PassManager* section, how can you write a Pass without deriving the `PassInfoMixin` class?

2. How can you develop a custom instrumentation for the new PassManager? Additionally, how can you do this without modifying the LLVM source tree? (Hint: think about the Pass plugin that we learned about in this chapter.)

10
Processing LLVM IR

In the previous chapter, we learned about PassManager and AnalysisManager in LLVM. We went through some tutorials for developing an LLVM pass and how to retrieve program analysis data via AnalysisManager. The knowledge and skills we acquired help to build the foundation for developers to create composable building blocks for code transformation and program analysis.

In this chapter, we are going to focus on the methodology of processing **LLVM IR**. LLVM IR is a target-independent **intermediate representation** for program analysis and compiler transformation. You can think of LLVM IR as an *alternative* form of the code you want to optimize and compile. However, different from the C/C++ code you're familiar with, LLVM IR describes the program in a different way – we will give you a more concrete idea later. The majority of the *magic* that's done by LLVM that makes the input program faster or smaller after the compilation process is performed on LLVM IR. Recall that in the previous chapter, *Chapter 9, Working with PassManager and AnalysisManager*, we described how different passes are organized in a pipeline fashion – that was the high-level structure of how LLVM transforms the input code. In this chapter, we are going to show you the fine-grained details of how to modify LLVM IR in an efficient way.

Although the most straightforward and visual way to view LLVM IR is by its textual representation, LLVM provides libraries that contain a set of powerful modern C++ APIs to interface with the IR. These APIs can inspect the in-memory representation of LLVM IR and help us manipulate it, which effectively changes the target program we are compiling. These LLVM IR libraries can be embedded in a wide variety of applications, allowing developers to transform and analyze their target source code with ease.

LLVM APIs for different programming languages

Officially, LLVM only supports APIs for two languages: C and C++. Between them, C++ is the most feature-complete and update to date, but it also has the most *unstable* interface – it might be changed at any time without backward compatibility. On the other hand, C APIs have stable interfaces but come at the cost of lagging behind new feature updates, or even keeping certain features absent. The API bindings for OCaml, Go, and Python are in the source tree as community-driven projects.

We will try to guide you with generally applicable learning blocks driven by commonly seen topics and tasks that are supported by many realistic examples. Here is the list of topics we'll cover in this chapter:

- Learning LLVM IR basics
- Working with values and instructions
- Working with loops

We will start by introducing LLVM IR. Then, we will learn about two of the most essential elements in LLVM IR – values and instructions. Finally, we'll end this chapter by looking at loops in LLVM – a more advanced topic that is crucial for working on performance-sensitive applications.

Technical requirements

The tools we'll need in this chapter are the `opt` command-line utility and `clang`. Please build them using the following command:

```
$ ninja opt clang
```

Most of the code in this chapter can be implemented inside LLVM pass – and the pass plugin – as introduced in *Chapter 9, Working with PassManager and AnalysisManager*.

In addition, please install the **Graphviz** tool. You can consult the following page for an installation guide for your system: `https://graphviz.org/download`. On Ubuntu, for instance, you can install that package via the following command:

```
$ sudo apt install graphviz
```

We will use a command-line tool – the `dot` command – provided by Graphviz to visualize the control flow of a function.

The code example mentioned in this chapter can be implemented inside LLVM pass, if not specified otherwise.

Learning LLVM IR basics

LLVM IR is an alternative form of the program you want to optimize and compile. It is, however, structured differently from normal programming languages such as C/C++. LLVM IR is organized in a hierarchical fashion. The levels in this hierarchy – counting from the top – are **Module**, **function**, **basic block**, and **instruction**. The following diagram shows their structure:

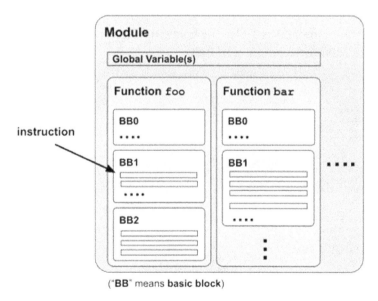

Figure 10.1 – Hierarchy structure of LLVM IR

A **module** represents a translation unit – usually a source file. Each module can contain multiple **functions** (or global variables). Each contains a list of **basic blocks** where each of the basic blocks contains a list of **instructions**.

> Quick refresher – basic block
>
> A basic block represents a list of instructions with only one entry and one exit point. In other words, if a basic block is executed, the control flow is guaranteed to walk through every instruction in the block.

Knowing the high-level structure of LLVM IR, let's look at one of the LLVM IR examples. Let's say we have the following C code, `foo.c`:

```c
int foo(int a, int b) {
   return a > 0? a - b : a + b;
}
```

We can use the following `clang` command to generate its *textual* LLVM IR counterpart:

```
$ clang -emit-llvm -S foo.c
```

The result will be put in the `foo.ll` file. The following diagram shows part of its content, with annotations for the corresponding IR unit:

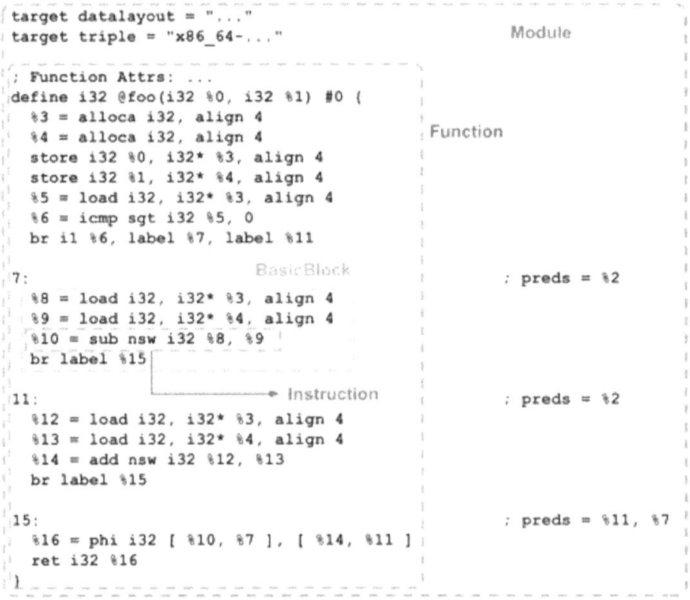

Figure 10.2 – Part of the content in foo.ll, annotated with the corresponding IR unit

In textual form, an instruction is usually presented in the following format:

```
<result> = <operator / op-code> <type>, [operand1, operand2, …]
```

For instance, let's assume we have the following instruction:

```
%12 = load i32, i32* %3
```

Here, `%12` is the result value, `load` is the op-code, `i32` is the data type of this instruction, and `%3` is the only operand.

In addition to the textual representation, nearly every component in LLVM IR has a C++ class counterpart with the same name. For instance, a function and a basic block are simply represented by `Function` and the `BasicBlock` C++ class, respectively.

Different kinds of instructions are represented by classes that are all derived from the `Instruction` class. For example, the `BinaryOperator` class represents a binary operation instruction, while `ReturnInst` class represents a return statement. We will look at `Instruction` and its child classes in more detail later.

The hierarchy depicted in *Figure 10.1* is the **concrete** structure of LLVM IR. That is, this is how they are stored in memory. On top of that, LLVM also provides other *logical* structures to view the relationships of different IR units. They are usually evaluated from the concrete structures and stored as auxiliary data structures or treated as analysis results. Here are some of the most important ones in LLVM:

- **Control Flow Graph** (**CFG**): This is a graph structure that's organized into basic blocks to show their control flow relations. The vertices in this graph represent basic blocks, while the edges represent a single control flow transfer.

- **Loop**: This represents the loop we are familiar with, which consists of multiple basic blocks that have at least one back edge – a control flow edge that goes back to its parent or ancestor vertices. We will look at this in more detail in the last section of this chapter, the *Working with loops* section.

- **Call graph**: Similar to CFG, the call graph also shows control flow transfers, but the vertices become individual functions and the edges become function call relations.

In the next section, we are going to learn how to iterate through different IR units in both concrete and logical structures.

Iterating different IR units

Iterating IR units – such as basic blocks or instructions – are *essential* to LLVM IR development. It is usually one of the first steps we must complete in many transformation or analysis algorithms – scanning through the whole code and finding an interesting area in order to apply certain measurements. In this section, we are going to learn about the practical aspects of iterating different IR units. We will cover the following topics:

- Iterating instructions
- Iterating basic blocks
- Iterating the call graph
- Learning the GraphTraits

Let's start by discussing how to iterate instructions.

Iterating instructions

An instruction is one of the most basic elements in LLVM IR. It usually represents a single action in the program, such as an arithmetic operation or a function call. Walking through all the instructions in a single basic block or function is the cornerstone of most program analyses and compiler optimizations.

To iterate through all the instructions in a basic block, you just need to use a simple for-each loop over the block:

```
// `BB` has the type of `BasicBlock&`
for (Instruction &I : BB) {
  // Work on `I`
}
```

We can iterate through all the instructions in a function in two ways. First, we can iterate over all the basic blocks in the function before visiting the instructions. Here is an example:

```
// `F` has the type of `Function&`
for (BasicBlock &BB : F) {
  for (Instruction &I : BB) {
    // Work on `I`
  }
}
```

Second, you can leverage a utility called `inst_iterator`. Here is an example:

```
#include "llvm/IR/InstIterator.h"
...
// `F` has the type of `Function&`
for (Instruction &I : instructions(F)) {
  // Work on `I`
}
```

Using the preceding code, you can retrieve all the instructions in this function.

Instruction visitor

There are many cases where we want to apply different treatments to different types of instructions in a basic block or function. For example, let's assume we have the following code:

```
for (Instruction &I : instructions(F)) {
  switch (I.getOpcode()) {
  case Instruction::BinaryOperator:
  // this instruction is a binary operator like `add` or `sub`
    break;
  case Instruction::Return:
    // this is a return instruction
    break;
    ...
  }
}
```

Recall that different kinds of instructions are modeled by (different) classes derived from `Instruction`. Therefore, an `Instruction` instance can represent any of them. The `getOpcode` method shown in the preceding snippet can give you a unique token – namely, `Instruction::BinaryOperator` and `Instruction::Return` in the given code – that tells you about the underlying class. However, if we want to work on the derived class (`ReturnInst`, in this case) instance rather than the "raw" `Instruction`, we need to do some type casting.

LLVM provides a better way to implement this kind of visiting pattern –InstVisitor. InstVisitor is a class where each of its member methods is a callback function for a specific instruction type. You can define your own callbacks after inheriting from the InstVisitor class. For instance, check out the following code snippet:

```
#include "llvm/IR/InstVisitor.h"
class MyInstVisitor : public InstVisitor<MyInstVisitor> {
  void visitBinaryOperator(BinaryOperator &BOp) {
    // Work on binary operator instruction
    …
  }
  void visitReturnInst(ReturnInst &RI) {
    // Work on return instruction
    …
  }
};
```

Each visitXXX method shown here is a callback function for a specific instruction type. Note that we are *not* overriding each of these methods (there was no override keyword attached to the method). Also, instead of defining callbacks for all the instruction types, InstVisitor allows you to only define those that we are interested in.

Once MyInstVisitor has been defined, we can simply create an instance of it and invoke the visit method to launch the visiting process. Let's take the following code as an example:

```
// `F` has the type of `Function&`
MyInstVisitor Visitor;
Visitor.visit(F);
```

There are also visit methods for Instruction, BasicBlock, and Module.

> **Ordering basic blocks and instructions**
>
> All the skills we've introduced in this section assume that the **ordering** of basic blocks or even instructions to visit is not your primary concern. However, it is important to know that Function doesn't store or iterate its enclosing BasicBlock instances in a particular *linear* order. We will show you how to iterate through all the basic blocks in various meaningful orders shortly.

With that, you've learned several ways to iterate instructions from a basic block or a function. Now, let's learn how to iterate basic blocks in a function.

Iterating basic blocks

In the previous section, we learned how to iterate basic blocks of a function using a simple for loop. However, developers can only receive basic blocks in an *arbitrary* order in this way – that ordering gave you neither the execution order nor the control flow information among the blocks. In this section, we will show you how to iterate basic blocks in a more meaningful way.

Basic blocks are important elements for expressing the control flow of a function, which can be represented by a directed graph – namely, the **CFG**. To give you a concrete idea of what a typical CFG looks like, we can leverage one of the features in the opt tool. Assuming you have an LLVM IR file, foo.ll, you can use the following command to print out the CFG of each function in Graphviz format:

```
$ opt -dot-cfg -disable-output foo.ll
```

This command will generate one .dot file for each function in foo.ll.

The .dot File might be hidden

The filename of the CFG .dot file for each function usually starts with a dot character ('.'). On Linux/Unix systems, this effectively *hides* the file from the normal ls command. So, use the ls -a command to show those files instead.

Each .dot file contains the Graphviz representation of that function's CFG. Graphviz is a general and textual format for expressing graphs. People usually convert a .dot file into other (image) formats before studying it. For instance, using the following command, you can convert a .dot file into a PNG image file that visually shows the graph:

```
$ dot -Tpng foo.cfg.dot > foo.cfg.png
```

The following diagram shows two examples:

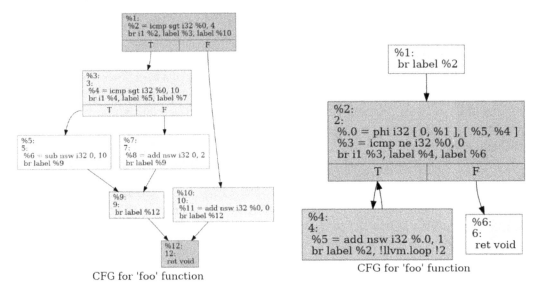

CFG for 'foo' function   CFG for 'foo' function

Figure 10.3 – Left: CFG for a function containing branches; right: CFG for a function containing a loop

The left-hand side of the preceding diagram shows a CFG for a function containing several branches; the right-hand side shows a CFG for a function containing a single loop.

Now, we know that basic blocks are organized as a directed graph – namely, the CFG. Can we iterate this CFG so that it follows the edges and nodes? LLVM answers this question by providing utilities for iterating a graph in four different ways: topological order, depth first (essentially doing **DFS**), breadth first (essentially doing **BFS**), and **Strongly Connected Components** (**SCCs**). We are going to learn how to use each of these utilities in the following subsections.

Let's start with topological order traversal.

Topological order traversal

Topological ordering is a simple linear ordering that guarantees that for each node in the graph, it will only be visited after we've visited all of its parent (predecessor) nodes. LLVM provides po_iterator and some other utility functions to implement *reversed* topological ordering (reversed topological ordering is easier to implement) on the CFG. The following snippet gives an example of using po_iterator:

```
#include "llvm/ADT/PostOrderIterator.h"
#include "llvm/IR/CFG.h"
// `F` has the type of `Function*`
```

```
for (BasicBlock *BB : post_order(F)) {
  BB->printAsOperand(errs());
  errs() << "\n";
}
```

The `post_order` function is just a helper function to create an iteration range of `po_iterator`. Note that the `llvm/IR/CFG.h` header is necessary to make `po_iterator` work on `Function` and `BasicBlock`.

If we apply the preceding code to the function containing branches in the preceding diagram, we'll get the following command-line output:

```
label %12
label %9
label %5
label %7
label %3
label %10
label %1
```

Alternatively, you can traverse from a specific basic block using nearly the same syntax; for instance:

```
// `F` has the type of `Function*`
BasicBlock &EntryBB = F->getEntryBlock();
for (BasicBlock *BB : post_order(&EntryBB)) {
  BB->printAsOperand(errs());
  errs() << "\n";
}
```

The preceding snippet will give you the same result as the previous one since it's traveling from the entry block. You're free to traverse from the arbitrary block, though.

Depth-first and breadth-first traversal

DFS and **BFS** are two of the most famous and iconic algorithms for visiting topological structures such as a graph or a tree. For each node in a tree or a graph, DFS will always try to visit its child nodes before visiting other nodes that share the same parents (that is, the *sibling* nodes). On the other hand, BFS will traverse all the sibling nodes before moving to its child nodes.

LLVM provides df_iterator and bf_iterator (and some other utility functions) to implement depth-first and breadth-first ordering, respectively. Since their usages are nearly identical, we are only going to demonstrate df_iterator here:

```
#include "llvm/ADT/DepthFirstIterator.h"
#include "llvm/IR/CFG.h"
// `F` has the type of `Function*`
for (BasicBlock *BB : depth_first(F)) {
  BB->printAsOperand(errs());
  errs() << "\n";
}
```

Similar to po_iterator and post_order, depth_first is just a utility function for creating an iteration range of df_iterator. To use bf_iterator, simply replace depth_first with breadth_first. If you apply the preceding code to the containing branches in the preceding diagram, it will give you the following command-line output:

```
label %1
label %3
label %5
label %9
label %12
label %7
label %10
```

When using bf_iterator/breadth_first, we will get the following command-line output for the same example:

```
label %1
label %3
label %10
label %5
label %7
label %12
label %9
```

df_iterator and bf_iterator can also be used with BasicBlock, in the same way as po_iterator shown previously.

SSC traversal

An **SCC** represents a subgraph where every enclosing node can be reached from every other node. In the context of CFG, it is useful to traverse CFG with loops.

The basic block traversal methods we introduced earlier are useful tools to reason about the control flow in a function. For a loop-free function, these methods give you a linear view that closely reflects the execution orders of enclosing basic blocks. However, for a function that contains loops, these (linear) traversal methods cannot show the cyclic execution flow that's created by the loops.

> **Recurring control flow**
>
> Loops are not the only programming constructs that create recurring control flows within a function. A few other directives – the `goto` syntax in C/C++, for example – will also introduce a recurring control flow. However, those corner cases will make analyzing the control flow more difficult (which is one of the reasons you shouldn't use `goto` in your code), so when we are talking about recurring control flows, we are only referring to loops.

Using `scc_iterator` in LLVM, we can traverse strongly connected basic blocks in a CFG. With this information, we can quickly spot a recurring control flow, which is crucial for some analysis and program transformation tasks. For example, we need to know the back edges and recurring basic blocks in order to accurately propagate the branch probability data along the control flow edges.

Here is an example of using `scc_iterator`:

```cpp
#include "llvm/ADT/SCCIterator.h"
#include "llvm/IR/CFG.h"
// `F` has the type of `Function*`
for (auto SCCI = scc_begin(&F); !SCCI.isAtEnd(); ++SCCI) {
  const std::vector<BasicBlock*> &SCC = *SCCI;
  for (auto *BB : SCC) {
    BB->printAsOperand(errs());
    errs() << "\n";
  }
  errs() << "====\n";
}
```

Different from the previous traversal methods, `scc_iterator` doesn't provide a handy range-style iteration. Instead, you need to create a `scc_iterator` instance using `scc_begin` and do manual increments. More importantly, you should use the `isAtEnd` method to check the exit condition, rather than doing a comparison with the "end" iterator like we usually do with C++ STL containers. A vector of `BasicBlock` can be dereferenced from a single `scc_iterator`. These `BasicBlock` instances are the basic blocks within a SCC. The ordering among these SCC instances is roughly the same as in the reversed topological order – namely, the post ordering we saw earlier.

If you run the preceding code over the function that contains a loop in the preceding diagram, it gives you the following command-line output:

```
label %6
====
label %4
label %2
====
label %1
====
```

This shows that basic blocks `%4` and `%2` are in the same SCC.

With that, you've learned how to iterate basic blocks within a function in different ways. In the next section, we are going to learn how to iterate functions within a module by following the call graph.

Iterating the call graph

A call graph is a direct graph that represents the function call relationships in a module. It plays an important role in **inter-procedural** code transformation and analysis, namely, analyzing or optimizing code across multiple functions. A famous optimization called **function inlining** is an example of this.

Before we dive into the details of iterating nodes in the call graph, let's take a look at how to build a call graph. LLVM uses the `CallGraph` class to represent the call graph of a single `Module`. The following sample code uses a pass module to build a `CallGraph`:

```
#include "llvm/Analysis/CallGraph.h"
struct SimpleIPO : public PassInfoMixin<SimpleIPO> {
  PreservedAnalyses run(Module &M, ModuleAnalysisManager &MAM)
{
    CallGraph CG(M);
```

```
    for (auto &Node : CG) {
      // Print function name
      if (Node.first)
        errs() << Node.first->getName() << "\n";
    }
    return PreservedAnalysis::all();
  }
};
```

This snippet built a `CallGraph` instance before iterating through all the enclosing functions and printing their names.

Just like `Module` and `Function`, `CallGraph` only provides the most basic way to enumerate all its enclosing components. So, how do we traverse `CallGraph` in different ways – for instance, by using SCC – as we saw in the previous section? The answer to this is surprisingly simple: in the exact *same* way – using the same set of APIs and the same usages.

The secret behind this is a thing called `GraphTraits`.

Learning about GraphTraits

`GraphTraits` is a class designed to provide an abstract interface over various different graphs in LLVM – CFG and call graph, to name a few. It allows other LLVM components – analyses, transformations, or iterator utilities, as we saw in the previous section – to build their works *independently* of the underlying graphs. Instead of asking every graph in LLVM to inherit from `GraphTraits` and implement the required functions, `GraphTraits` takes quite a different approach by using **template specialization**.

Let's say that you have written a simple C++ class that has a template argument that accepts arbitrary types, as shown here:

```
template <typename T>
struct Distance {
  static T compute(T &PointA, T &PointB) {
    return PointA - PointB;
  }
};
```

This C++ class will compute the distance between two points upon calling the `Distance::compute` method. The types of those points are parameterized by the `T` template argument.

If `T` is a numeric type such as `int` or `float`, everything will be fine. However, if `T` is a struct of a class, like the one here, then the default `compute` method implementation will not be able to compile:

```
Distance<int>::compute(94, 87); // Success
...
struct SimplePoint {
  float X, Y;
};
SimplePoint A, B;
Distance<SimplePoint>::compute(A, B); // Compilation Error
```

To solve this issue, you can either implement a subtract operator for `SimplePoint`, or you can use template specialization, as shown here:

```
// After the original declaration of struct Distance…
template<>
struct Distance<SimplePoint> {
  SimplePoint compute(SimplePoint &A, SimplePoint &B) {
    return std::sqrt(std::pow(A.X - B.X, 2),…);
  }
};
...
SimplePoint A, B;
Distance<SimplePoint>::compute(A, B); // Success
```

`Distance<SimplePoint>` in the previous code describes what `Distance<T>` looks like when `T` is equal to `SimplePoint`. You can think of the original `Distance<T>` as some kind of **interface** and `Distance<SimplePoint>` being one of its **implementations**. But be aware that the `compute` method in `Distance<SimplePoint>` is *not* an override method of the original `compute` in `Distance<T>`. This is different from normal class inheritance (and virtual methods).

GraphTraits in LLVM is a template class that provides an interface for various graph algorithms, such as df_iterator and scc_iterator, as we saw previously. Every graph in LLVM will *implement* this interface via template specialization. For instance, the following GraphTraits specialization is used for modeling the **CFG** of a function:

```
template<>
struct GraphTraits<Function*> {…}
```

Inside the body of GraphTraits<Function*>, there are several (static) methods and typedef statements that implement the required interface. For example, nodes_iterator is the type that's used for iterating over all the vertices in CFG, while nodes_begin provides you with the entry/starting node of this CFG:

```
template<>
struct GraphTraits<Function*> {
  typedef pointer_iterator<Function::iterator> nodes_iterator;
  static node_iterator nodes_begin(Function *F) {
    return nodes_iterator(F->begin());
  }
  …
};
```

In this case, nodes_iterator is basically Function::iterator. nodes_begin simply returns the first basic block in the function (via an iterator). If we look at GraphTraits for CallGraph, it has completely different implementations of nodes_iterator and nodes_begin:

```
template<>
struct GraphTraits<CallGraph*> {
  typedef mapped_iterator<CallGraph::iterator,
  decltype(&CGGetValuePtr)> nodes_iterator;
  static node_iterator nodes_begin(CallGraph *CG) {
    return nodes_iterator(CG->begin(), &CGGetValuePtr);
  }
};
```

When developers are implementing a new graph algorithm, instead of hardcoding it for each kind of graph in LLVM, they can build their algorithms by using GraphTraits as an interface to access the key properties of arbitrary graphs.

For example, let's say we want to create a new graph algorithm, `find_tail`, which finds the first node in the graph that has no child nodes. Here is the skeleton of `find_tail`:

```
template<class GraphTy,
         typename GT = GraphTraits<GraphTy>>
  auto find_tail(GraphTy G) {
  for(auto NI = GT::nodes_begin(G); NI != GT::nodes_end(G);
   ++NI) {
    // A node in this graph
    auto Node = *NI;
    // Child iterator for this particular node
    auto ChildIt = GT::child_begin(Node);
    auto ChildItEnd = GT::child_end(Node);
    if (ChildIt == ChildItEnd)
      // No child nodes
      return Node;
  }
  ...
}
```

With the help of this template and `GraphTraits`, we can *reuse* this function on `Function`, `CallGraph`, or any kind of graph in LLVM; for instance:

```
// `F` has the type of `Function*`
BasicBlock *TailBB = find_tail(F);
// `CG` has the type of `CallGraph*`
CallGraphNode *TailCGN = find_tail(CG);
```

In short, `GraphTraits` generalizes algorithms – such as `df_iterator` and `scc_iterator`, as we saw previously – in LLVM to *arbitrary* graphs using the template specialization technique. This is a clean and efficient way to define interfaces for reusable components.

In this section, we learned the hierarchy structure of LLVM IR and how to iterate different IR units – either concrete or logical units, such as CFGs. We also learned the important role of `GraphTraits` for encapsulating different graphs – CFGs and call graphs, to name a few – and exposed a common interface to various algorithms in LLVM, thus making those algorithms more concise and reusable.

In the next section, we will learn about how values are represented in LLVM, which describes a picture of how different LLVM IR components are associated with each other. In addition, we will learn about the correct and efficient way to manipulate and update values in LLVM.

Working with values and instructions

In LLVM, a **value** is a unique construct – not only does it represent values stored in variables, but it also models a wide range of concepts from constants, global variables, individual instructions, and even basic blocks. In other words, it is one of the *foundations* of LLVM IR.

The concept of value is especially important for instructions as it directly interacts with values in the IR. Therefore, in this section, we will put them into the same discussion. We are going to see how values work in LLVM IR and how values are associated with instructions. On top of that, we are going to learn how to create and insert new instructions, as well as how to update them.

To learn how to use values in LLVM IR, we must understand the important theory behind this system, which dictates the behavior and the format of LLVM instructions – the **Single Static Assignment** (**SSA**) form.

Understanding SSA

SSA is a way of structuring and designing IR to make program analysis and compiler transformation easier to perform. In SSA, a variable (in the IR) will only be assigned a value exactly *once*. This means that we cannot manipulate a variable like this:

```
// the following code is NOT in SSA form
x = 94;
x = 87; // `x` is assigned the second time, not SSA!
```

Although a variable can only be assigned once, it can be *used* multiple times in arbitrary instructions. For instance, check out the following code:

```
x = 94;
y = x + 4; // first time `x` is used
z = x + 2; // second time `x` is used
```

You might be wondering how normal C/C++ code – which is clearly not in SSA form – gets transformed into an SSA form of IR, such as LLVM. While there is a whole class of different algorithms and research papers that answer this question, which we are not going to cover here, most of the simple C/C++ code can be transformed using trivial techniques such as renaming. For instance, let's say we have the following (non-SSA) C code:

```
x = 94;
x = x * y; // `x` is assigned more than once, not SSA!
x = x + 5;
```

Here, we can rename x in the first assignment with something like x0 and x on the left-hand side of the second and third assignments with alternative names such as x1 and x2, respectively:

```
x0 = 94;
x1 = x0 * y;
x2 = x1 + 5;
```

With these simple measurements, we can obtain the SSA form of our original code with the same behavior.

To have a more comprehensive understanding of SSA, we must change our way of thinking about what instructions *look like* in a program. In **imperative programming languages** such as C/C++, we often treat each statement (instruction) as an **action**. For instance, in the following diagram, on the left-hand side, the first line represents an action that "assigns 94 to variable x" where the second line means "do some multiplication using x and y before storing the result in the x variable":

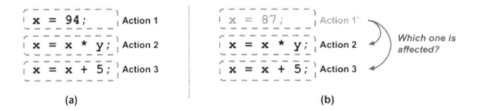

Figure 10.4 – Thinking instructions as "actions"

These interpretations sound intuitive. However, things get tricky when we make some *transformations* – which is, of course, a common thing in a compiler – on these instructions. In the preceding diagram, on the right-hand side, when the first instruction becomes x = 87, we don't know if this modification **affects** other instructions. If it does, we are not sure which one of these gets affected. This information not only tells us whether there are other potential opportunities to optimize, but it is also a crucial factor for the *correctness* of compiler transformation – after all, no one wants a compiler that will break your code when optimization is enabled. What's worse, let's say we are inspecting the x variable on the right-hand side of **Action 3**. We are interested in the last few instructions that modify this x. Here, we have no choice but to list all the instructions that have x on their left-hand side (that is, using x as the destination), which is pretty inefficient.

Instead of looking at the *action* aspect of an instruction, we can focus on the *data* that's generated by an instruction and get a clear picture of the provenance of each instruction – that is, the region that can be reached by its resulting value. Furthermore, we can easily find out the origins of arbitrary variables/values. The following diagram illustrates this advantage:

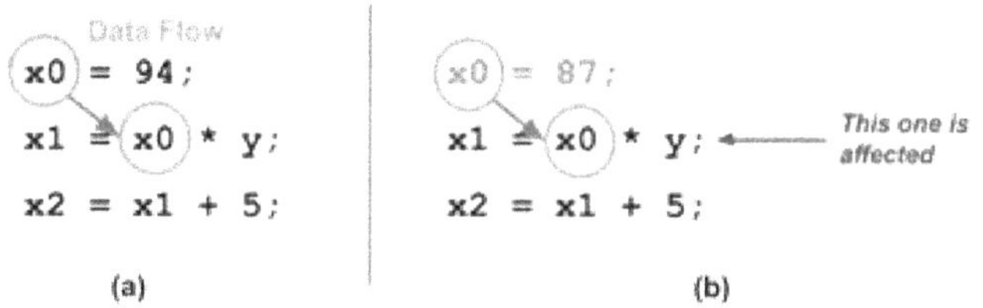

Figure 10.5 – SSA highlighting the dataflow among instructions

In other words, SSA highlights the **dataflow** in a program so that the compiler will have an easier time tracking, analyzing, and modify the instructions.

Instructions in LLVM are organized in SSA form. This means we are more interested in the value, or the data flow generated by an instruction, rather than which variable it stores the result in. Since each instruction in LLVM IR can only produce a single result value, an Instruction object – recall that Instruction is the C++ class that represents an instruction in LLVM IR – also represents its **result value**. To be more specific, the concept of *value* in LLVM IR is represented by a C++ class called Value. Instruction is one of its child classes. This means that given an Instruction object, we can, of course, cast it to a Value object. That particular Value object is effectively the result of that Instruction:

```
// let's say `I` represents an instruction `x = a + b`
Instruction *I = …;
Value *V = I; // `V` effectively represents the value `x`
```

This is one of the most important things to know in order to work with LLVM IR, especially to use most of its APIs.

While the `Instruction` object represents its own result value, it also has *operands* that are served as inputs to the instruction. Guess what? We are also using `Value` objects as operands. For example, let's assume we have the following code:

```
Instruction *BinI = BinaryOperator::Create(Instruction::Add,…);
Instruction *RetI = ReturnInst::Create(…, BinI, …);
```

The preceding snippet is basically creating an arithmetic addition instruction (represented by `BinaryOperator`), whose *result* value will be the *operand* of another return instruction. The resulting IR is equivalent to the following C/C++ code:

```
x = a + b;
return x;
```

In addition to `Instruction`, `Constant` (the C++ class for different kinds of constant values), `GlobalVariable` (the C++ class for global variables), and `BasicBlock` are all subclasses of `Value`. This means that they're also organized in SSA form and that you can use them as the operands for an `Instruction`.

Now, you know what SSA is and learned what impact it has on the design of LLVM IR. In the next section, we are going to discuss how to modify and update values in LLVM IR.

Working with values

SSA makes us focus on the *data flow* among instructions. Since we have a clear view of how values go from one instruction to the other, it's easy to replace the usage of certain values in an instruction. But how is the concept of "value usage" represented in LLVM? The following diagram shows two important C++ classes that answer this question – `User` and `Use`:

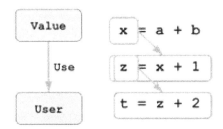

Figure 10.6 – The relationship between Value, User, and Use

As we can see, `User` represents the concept of an IR instance (for example, an `Instruction`) that uses a certain `Value`. Furthermore, LLVM uses another class, `Use`, to model the edge between `Value` and `User`. Recall that `Instruction` is a child class of `Value` – which represents the result that's generated by this instruction. In fact, `Instruction` is also derived from `User`, since almost all instructions take at least one operand.

A `User` might be pointed to by multiple `Use` instances, which means it uses many `Value` instances. You can use `value_op_iterator` provided by `User` to check out each of these `Value` instances; for example:

```
// `Usr` has the type of `User*`
for (Value *V : Usr->operand_values()) {
    // Working with `V`
}
```

Again, `operand_values` is just a utility function to generate a `value_op_iterator` range.

Here is an example of why we want to iterate through all the `User` instances of a `Value`: imagine we are analyzing a program where one of its `Function` instances will return sensitive information – let's say, a `get_password` function. Our goal is to ensure that whenever `get_password` is called within a `Function`, its returned value (sensitive information) won't be leaked via another function call. For example, we want to detect the following pattern and raise an alarm:

```
void vulnerable() {
    v = get_password();
    …
    bar(v); // WARNING: sensitive information leak to `bar`!
}
```

One of the most naïve ways to implement this analysis is by inspecting all `User` instances of the sensitive `Value`. Here is some example code:

```
User *find_leakage(CallInst *GetPWDCall) {
    for (auto *Usr : GetPWDCall->users()) {
        if (isa<CallInst>(Usr)) {
            return Usr;
        }
```

```
    }
  …
}
```

The find_leackage function takes a CallInst argument – which represents a get_ password function call – and returns any User instance that uses the Value instance that's returned from that get_password call.

A Value instance can be used by multiple different User instances. So, similarly, we can iterate through all of them using the following snippet:

```
// `V` has the type of `Value*`
for (User *Usr : V->users()) {
  // Working with `Usr`
}
```

With that, you've learned how to inspect the User instance of a Value, or the Value instance that's used by the current User. In addition, when we're developing a compiler transformation, it is pretty common to change the Value instance that's used by a User to another one. LLVM provides some handy utilities to do this.

First, the Value::replaceAllUsesWith method can, as its name suggests, tell all of its User instances to use another Value instead of it. The following diagram illustrates its effect:

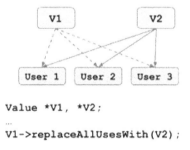

```
Value *V1, *V2;
…
V1->replaceAllUsesWith(V2);
```

Figure 10.7 – Effect of Value::replaceAllUsesWith

This method is really useful when you're replacing an Instruction with another Instruction. Using the preceding diagram to explain this, V1 is the original Instruction and V2 is the new one.

Another utility function that does a similar thing is
`User::replaceUsesOfWith(From,To)`. This method effectively scans through
all of the operands in this `User` and replaces the usage of a specific `Value` (the *From*
argument) with another `Value` (the *To* argument).

The skills you've learned in this section are some of the most fundamental tools for
developing a program transformation in LLVM. In the next section, we will talk about
how to create and modify instructions.

Working with instructions

Previously, we learned the basics of `Value` – including its relationship with
`Instruction` – and the way to update `Value` instances under the framework of SSA.
In this section, we are going to learn some more basic knowledge and skills that give you
a better understanding of `Instruction` and help you *modify* `Instruction` instances
in a correct and efficient way, which is the key to developing a successful compiler
optimization.

Here is the list of topics we are going to cover in this section:

- Casting between different instruction types
- Inserting a new instruction
- Replacing an instruction
- Processing instructions in batches

Let's start by looking at different instruction types.

Casting between different instruction types

In the previous section, we learned about a useful utility called `InstVisitor`. The
`InstVisitor` class helps you determine the underlying class of an `Instruction`
instance. It also saves you the efforts of casting between different instruction types.
However, we cannot always rely on `InstVisitor` for every task that involves type
casting between `Instruction` and its derived classes. More generally speaking, we want
a simpler solution for type casting between parent and child classes.

Now, you might be wondering, but C++ *already* provided this mechanism via the
`dynamic_cast` directive, right? Here is an example of `dynamic_cast`:

```
class Parent {…};
class Child1 : public Parent {…};
class Child2 : public Parent {…};
```

```
void foo() {
  Parent *P = new Child1();
  Child1 *C = dynamic_cast<Child1*>(P); // OK
  Child2 *O = dynamic_cast<Child2*>(P); // Error: bails out at
                                        // runtime
}
```

In the `foo` function used in the preceding code, we can see that in its second line, we can convert P into a `Child1` instance because that is its underlying type. On the other hand, we cannot convert P into `Child2` – the program will simply crash during runtime if we do so.

Indeed, `dynamic_cast` has the exact functionality we are looking for – more formally speaking, the **Runtime Type Info** (**RTTI**) feature – but it also comes with high overhead in terms of runtime performance. What's worse, the default implementation of RTTI in C++ is quite complex, making the resulting program difficult to optimize. Therefore, LLVM *disables* RTTI by default. Due to this, LLVM came up with its own system of runtime type casting that is much simpler and more efficient. In this section, we are going to talk about how to use it.

LLVM's casting framework provides three functions for dynamic type casting:

- `isa<T>(val)`
- `cast<T>(val)`
- `dyn_cast<T>(val)`

The first function, `isa<T>` – pronounced "is-a" – checks if the `val` pointer type can be cast to a pointer of the `T` type. Here is an example:

```
// `I` has the type of `Instruction*`
if (isa<BinaryOperator>(I)) {
  // `I` can be casted to `BinaryOperator*`
}
```

Note that differently from `dynamic_cast`, you don't need to put `BinaryOperator*` as the template argument in this case – only a type without a pointer qualifier.

The `cast<T>` function performs the real type casting from (pointer-type) `val` to a pointer of the `T` type. Here is an example:

```
// `I` has the type of `Instruction*`
if (isa<BinaryOperator>(I)) {
  BinaryOperator *BinOp = cast<BinaryOperator>(I);
}
```

Again, you don't need to put `BinaryOperator*` as the template argument. Note that if you don't perform type checking using `isa<T>` before calling `cast<T>`, the program will just crash during runtime.

The last function, `dyn_cast<T>`, is a combination of `isa<T>` and `cast<T>`; that is, you perform type casting if applicable. Otherwise, it returns a null. Here is an example:

```
// `I` has the type of `Instruction*`
if (BinaryOperator *BinOp = dyn_cast<BinaryOperator>(I)) {
  // Work with `BinOp`
}
```

Here, we can see some neat syntax that combines the variable declaration (of `BinOp`) with the `if` statement.

Be aware that none of these APIs can take null as the argument. On the contrary, `dyn_cast_or_null<T>` doesn't have this limitation. It is basically a `dyn_cast<T>` API that accepts null as input.

Now, you know how to check and cast from an arbitrary `Instruction` instance to its underlying instruction type. Starting from the next section, we are finally going to create and modify some instructions.

Inserting a new instruction

In one of the code examples from the previous *Understanding SSA* section, we saw a snippet like this:

```
Instruction *BinI = BinaryOperator::Create(…);
Instruction *RetI = ReturnInst::Create(…, BinI, …);
```

As suggested by the method's name – `Create` – we can infer that these two lines created a `BinaryOperator` and a `ReturnInst` instruction.

Most of the instruction classes in LLVM provide factory methods – such as `Create` here – to build a new instance. People are encouraged to use these factory methods versus allocating instruction objects manually via the `new` keyword or `malloc` function. LLVM will manage the instruction object's memory for you – once it's been inserted into a `BasicBlock`. There are several ways to insert a new instruction into a `BasicBlock`:

- Factory methods in some instruction classes provide an option to insert the instruction right after it is created. For instance, one of the `Create` method variants in `BinaryOperator` allows you to insert it *before* another instruction after the creation. Here is an example:

```
Instruction *BeforeI = …;
auto *BinOp = BinaryOperator::Create(Opcode, LHS, RHS,
  "new_bin_op", BeforeI);
```

In such a case, the instruction represented by `BinOp` will be placed before the one represented by `BeforeI`. This method, however, can't be ported across different instruction classes. Not every instruction class has factory methods that provide this feature and even if they do provide them, the API might not be the same.

- We can use the `insertBefore`/`insertAfter` methods provided by the `Instruction` class to insert a new instruction. Since all instruction classes are subclasses of `Instruction`, we can use `insertBefore` or `insertAfter` to insert the newly created instruction instance before or after another `Instruction`.

- We can also use the `IRBuilder` class. `IRBuilder` is a powerful tool for automating some of the instruction creation and insertion steps. It implements a builder design pattern that can insert new instructions one after another when developers invoke one of its creation methods. Here is an example:

```
// `BB` has the type of `BasicBlock*`
IRBuilder<> Builder(BB /*the insertion point*/);
// insert a new addition instruction at the end of `BB`
auto *AddI = Builder.CreateAdd(LHS, RHS);
// Create a new `ReturnInst`, which returns the result
// of `AddI`, and insert after `AddI`
Builer.CreateRet(AddI);
```

First, when we create an IRBuilder instance, we need to designate an *insertion point* as one of the constructor arguments. This insertion point argument can be a BasicBlock, which means we want to insert a new instruction at the end of BasicBlock; it can also be an Instruction instance, which means that new instructions are going to be inserted *before* that specific Instruction.

You are encouraged to use IRBuilder over other mechanisms if possible whenever you need to create and insert new instructions in sequential order.

With that, you've learned how to create and insert new instructions. Now, let's look at how to *replace* existing instructions with others.

Replacing an instruction

There are many cases where we will want to replace an existing instruction. For instance, a simple optimizer might replace an arithmetic multiplication instruction with a left-shifting instruction when one of the multiplication's operands is a power-of-two integer constant. In this case, it seems straightforward that we can achieve this by simply changing the *operator* (the opcode) and one of the operands in the original Instruction. That is **not** the recommended way to do things, however.

To replace an Instruction in LLVM, you need to create a new Instruction (as the replacement) and *reroute* all the SSA definitions and usages from the original Instruction to the replacement one. Let's use the power-of-two-multiplication we just saw as an example:

1. The function we are going to implement is called replacePow2Mul, whose argument is the multiplication instruction to be processed (assuming that we have ensured the multiplication has a constant, power-of-two integer operand). First, we will retrieve the constant integer – represented by the ConstantInt class – operand and convert it into its base-2 logarithm value (via the getLog2 utility function; the exact implementation of getLog2 is left as an exercise for you):

    ```
    void replacePow2Mul(BinaryOperator &Mul) {
      // Find the operand that is a power-of-2 integer
      // constant
      int ConstIdx = isa<ConstantInt>(Mul.getOperand(0))? 0
        : 1;
      ConstantInt *ShiftAmount = getLog2(Mul.
        getOperand(ConstIdx));
    }
    ```

2. Next, we will create a new left-shifting instruction – represented by the `ShlOperator` class:

```
void replacePow2Mul(BinaryOperator &Mul) {
    …
    // Get the other operand from the original instruction
    auto *Base = Mul.getOperand(ConstIdx? 0 : 1);
    // Create an instruction representing left-shifting
    IRBuilder<> Builder(&Mul);
    auto *Shl = Builder.CreateShl(Base, ShiftAmount);
}
```

3. Finally, before we remove the `Mul` instruction, we need to tell all the users of the original `Mul` to use our newly created `Shl` instead:

```
void replacePow2Mul(BinaryOperator &Mul) {
    …
    // Using `replaceAllUsesWith` to update users of `Mul`
    Mul.replaceAllUsesWith(Shl);
    Mul.eraseFromParent(); // remove the original
                           // instruction
}
```

Now, all the original users of `Mul` are using `Shl` instead. Thus, we can safely remove `Mul` from the program.

With that, you've learned how to replace an existing `Instruction` properly. In the final subsection, we are going to talk about some tips for processing multiple instructions in a `BasicBlock` or a `Function`.

Tips for processing instructions in batches

So far, we have been learning how to insert, delete, and replace a single `Instruction`. However, in real-world cases, we usually perform such actions on a *sequence* of `Instruction` instances (that are in a `BasicBlock`, for instance). Let's try to do that by putting what we've learned into a `for` loop that iterates through all the instructions in a `BasicBlock`; for instance:

```
// `BB` has the type of `BasicBlock&`
for (Instruction &I : BB) {
  if (auto *BinOp = dyn_cast<BinaryOperator>(&I)) {
```

```
    if (isMulWithPowerOf2(BinOp))
      replacePow2Mul(BinOp);
  }
}
```

The preceding code used the `replacePow2Mul` function we just saw in the previous section to replace the multiplications in this `BasicBlock` with left-shifting instructions if the multiplication fulfills certain criteria. (This is checked by the `isMulWithPowerOf2` function. Again, the details of this function have been left as an exercise to you.)

This code looks pretty straightforward but unfortunately, it will crash while running this transformation. What happened here was that the **iterator** that's used for enumerating `Instruction` instances in `BasicBlock` became *stale* after running our `replacePow2Mul`. The `Instruction` iterator is unable to keep updated with the changes that have been applied to the `Instruction` instances in this `BasicBlock`. In other words, it's really hard to change the `Instruction` instances while iterating them at the same time.

The simplest way to solve this problem is to **push off** the changes:

```
// `BB` has the type of `BasicBlock&`
std::vector<BinaryOperator*> Worklist;
// Only perform the feasibility check
for (auto &I : BB) {
  if (auto *BinOp = dyn_cast<BinaryOperator>(&I)) {
    if (isMulWithPowerOf2(BinOp)) Worklist.push_back(BinOp);
  }
}
// Replace the target instructions at once
for (auto *BinOp : Worklist) {
  replacePow2Mul(BinOp);
}
```

The preceding code separates the previous code example into two parts (as two separate `for` loops). The first `for` loop is still iterating through all the `Instruction` instances in `BasicBlock`. But this time, it only performs the checks (that is, calling `isMulWithPowerOf2`) without replacing the `Instruction` instance right away if it passes the checks. Instead, this `for` loop pushes the candidate `Instruction` into array storage – a **worklist**. After finishing the first `for` loop, the second `for` loop inspects the worklist and performs the real replacement by calling `replacePow2Mul` on each worklist item. Since the replacements in the second `for` loop don't invalidate any iterators, we can finally transform the code without any crashes occurring.

There are, of course, other ways to circumvent the aforementioned iterator problem, but they are mostly complicated and less readable. Using a worklist is the safest and most expressive way to modify instructions in batches.

`Value` is a first-class construction in LLVM that outlines the data flow among different entities such as instructions. In this section, we introduced how values are represented in LLVM IR and the model of SSA that makes it easier to analyze and transform it. We also learned how to update values in an efficient way and some useful skills for manipulating instructions. This will help build the foundation for you to build more complex and advanced compiler optimizations using LLVM.

In the next section, we will look at a slightly more complicated IR unit – a loop. We are going to learn how loops are represented in LLVM IR and how to work with them.

Working with loops

So far, we have learned about several IR units such as modules, functions, basic blocks, and instructions. We have also learned about some *logical* units such as CFG and call graphs. In this section, we are going to look at a more logical IR unit: a loop.

Loops are ubiquitous constructions that are heavily used by programmers. Not to mention that nearly every programming language contains this concept, too. A loop repeatedly executes a certain number of instructions multiple times, which, of course, saves programmers lots of effort from repeating that code by themselves. However, if the loop contains any *inefficient* code – for example, a time-consuming memory load that always delivers the same value – the performance slowdown will also be *magnified* by the number of iterations.

Therefore, it is the compiler's job to eliminate as many flaws as possible from a loop. In addition to removing suboptimal code from loops, since loops are on the critical path of the runtime's performance, people have always been trying to further optimize them with special hardware-based accelerations; for example, replacing a loop with vector instructions, which can process multiple scalar values in just a few cycles. In short, loop optimization is the key to generating faster, more efficient programs. This is especially important in the high-performance and scientific computing communities.

In this section, we are going to learn how to process loops with LLVM. We will try to tackle this topic in two parts:

- Learning about loop representation in LLVM
- Learning about loop infrastructure in LLVM

In LLVM, loops are slightly more complicated than other (logical) IR units. Therefore, we will learn about the high-level concept of a loop in LLVM and its *terminologies* first. Then, in the second part, we are going to get our hands on the infrastructure and tools that are used for processing loops in LLVM.

Let's start with the first part.

Learning about loop representation in LLVM

A loop is represented by the `Loop` class in LLVM. This class captures any control flow structure that has a *back edge* from an enclosing basic block in one of its predecessor blocks. Before we dive into its details, let's learn how to retrieve a `Loop` instance.

As we mentioned previously, a loop is a logical IR unit in LLVM IR. Namely, it is derived (or calculated) from physical IR units. In this case, we need to retrieve the calculated `Loop` instances from `AnalysisManager` – which was first introduced in *Chapter 9, Working with PassManager and AnalysisManager*. Here is an example showing how to retrieve it in a `Pass` function:

```
#include "llvm/Analysis/LoopInfo.h"
...
PreservedAnalyses run(Function &F, FunctionAnalysisManager
&FAM) {
  LoopInfo &LI = FAM.getResult<LoopAnalysis>(F);
  // `LI` contains ALL `Loop` instances in `F`
  for (Loop *LP : LI) {
    // Working with one of the loops, `LP`
```

```
    }
    ...
}
```

`LoopAnalysis` is an LLVM analysis class that provides us with a `LoopInfo` instance, which includes *all* the `Loop` instances in a `Function`. We can iterate through a `LoopInfo` instance to get an individual `Loop` instance, as shown in the preceding code.

Now, let's look into a `Loop` instance.

Learning about loop terminologies

A `Loop` instance contains multiple `BasicBlock` instances for a particular loop. LLVM assigns a special meaning/name to some of these blocks, as well as the (control flow) edges among them. The following diagram shows this terminology:

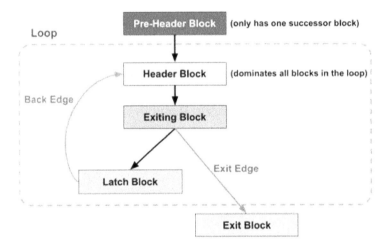

Figure 10.8 – Structure and terminology used in a loop

Here, every rectangle is a `BasicBlock` instance. However, only blocks residing within the dash line area are included in a `Loop` instance. The preceding diagram also shows two important control flow edges. Let's explain each of these terminologies in detail:

- **Header Block**: This block marks the *entry* to a loop. Formally speaking, it dominates all the blocks in the loop. It is also the destination of any back edge.

  Pre-Header Block: Though it is *not* part of a `Loop`, it represents the block that has the header block as the only successor. In other words, it's the only predecessor of the header block.

The existence of a pre-header block makes it easier to write some of the loop transformations. For instance, when we want to *hoist* an instruction to the outside of the loop so that it is only executed once before entering the loop, the pre-header block can be a good place to put this. If we don't have a pre-header block, we need to duplicate this instruction for *every* predecessor of the header block.

- **Back Edge**: This is the control flow edge that goes from one of the blocks in the loop to the header block. A loop might contain several back edges.

- **Latch Block**: This is the block that sits at the source of a back edge.

- **Exiting Block** and **Exit Block**: These two names are slightly confusing: the exiting block is the block that has a control flow edge – the **Exit Edge** – that goes *outside* the loop. The other end of the exit edge, which is not part of the loop, is the exit block. A loop can contain multiple exit blocks (and exiting blocks).

These are the important terminologies for blocks in a `Loop` instance. In addition to the control flow structure, compiler engineers are also interested in a special value that might exist in a loop: the **induction variable**. For example, in the following snippet, the `i` variable is the induction variable:

```
for (int i = 0; i < 87; ++i) {…}
```

A loop might not contain an induction variable – for example, many `while` loops in C/C++ don't have one. Also, it's not always easy to find out about an induction variable, nor its *boundary* – the start, end, and stopping values. We will show some of the utilities in the next section to help you with this task. But before that, we are going to discuss an interesting topic regarding the *canonical* form of a loop.

Understanding canonical loops

In the previous section, we learned several pieces of terminology for loops in LLVM, including the pre-header block. Recall that the existence of a pre-header block makes it easier to develop a loop transformation because it creates a simpler loop structure. Following this discussion, there are other properties that make it easier for us to write loop transformations, too. If a `Loop` instance has these nice properties, we usually call it a **canonical loop**. The optimization pipeline in LLVM will try to "massage" a loop into this canonical form before sending it to any of the loop transformations.

Currently, LLVM has two canonical forms for `Loop`: a **simplified** form and a **rotated** form. The simplified form has the following properties:

- A pre-header block.

- A single back edge (and thus a single latch block).

- The predecessors of the exit blocks come from the loop. In other words, the header block dominates all the exit blocks.

To get a simplified loop, you can run `LoopSimplfyPass` over the original loop. In addition, you can use the `Loop::isLoopSimplifyForm` method to check if a `Loop` is in this form.

The benefits of having a single back edge include that we can analyze recursive data flow – for instance, the induction variable – more easily. For the last property, if every exit block is dominated by the loop, we can have an easier time "sinking" instructions below the loop without any interference from other control flow paths.

Let's look at the rotated canonical form. Originally, the rotated form was not a formal canonical form in LLVM's loop optimization pipeline. But with more and more loop passes depending on it, it has become the "de facto" canonical form. The following diagram shows what this form looks like:

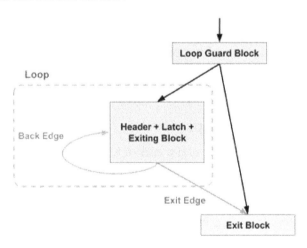

Figure 10.9 – Structure and terminology of a rotated loop

To get a rotated loop, you can run `LoopRotationPass` over the original loop. To check if a loop is rotated, you can use the `Loop::isRotatedForm` method.

This rotated form is basically transforming an arbitrary loop into a `do{…}while(…)` loop (in C/C++) with some extra checks. More specifically, let's say we have the following `for` loop:

```
// `N` is not a constant
for (int i = 0; i < N; ++i){…}
```

Loop rotation effectively turns it into the following code:

```
if (i < N) {
   do {

      …

      ++i;
   } while(i < N);
}
```

The highlighted boundary check in the preceding code is used to ensure that the loop won't execute if the `i` variable is out of bounds at the very beginning. We also call this check the **loop guard**, as shown in the preceding diagram.

In addition to the loop guard, we also found that a rotated loop has a combined header, latch, and exiting block. The rationale behind this is to ensure that every instruction in this block has the same *execution count*. This is a useful property for compiler optimizations such as loop vectorization.

With that, we have learned about the various loop terminologies and the definition of canonical loops in LLVM. In the next section, we will learn about some APIs that can help you inspect some of these properties and process loops in an efficient way.

Learning about loop infrastructure in LLVM

In the *Learning about loop representation in LLVM* section, we learned about the high-level construction and important properties of a loop in LLVM IR. In this section, we are going to see what APIs are available for us to inspect those properties and further transform the loops. Let's start our discussion from the loop pass – the LLVM pass that applies to `Loop` instances.

In *Chapter 9, Working with PassManager and AnalysisManager*, we learned that there are different kinds of LLVM pass that work on different IR units – for instance, we have seen passes for `Function` and `Module`. These two kinds of passes have a similar function signature for their `run` method – the main entry point of an LLVM pass – as shown here:

```
PreservedAnalyses run(<IR unit class> &Unit,
                      <IR unit>AnalysisManager &AM);
```

Both of their `run` methods take two arguments – a reference to the IR unit instance and an `AnalysisManager` instance.

In contrast, a loop pass has a slightly more complicated `run` method signature, as shown here:

```
PreservedAnalyses run(Loop &LP, LoopAnalysisManager &LAM,
                      LoopStandardAnalysisResults &LAR,
                      LPMUpdater &U);
```

The `run` method takes four arguments, but we already know about the first two. Here are the descriptions for the other two:

- The third argument, `LoopStandardAnalysisResults`, provides you with some analysis data instances, such as `AAResults` (alias analysis data), `DominatorTree`, and `LoopInfo`. These analyses are extensively used by many loop optimizations. However, most of them are managed by either `FunctionAnalysisManager` or `ModuleAnalysisManager`. This means that, originally, developers needed to implement more complicated methods – for example, using the `OuterAnalysisManagerProxy` class – to retrieve them. The `LoopStandardAnalysisResults` instance basically helps you retrieve this analysis data *ahead of time*.

- The last argument is used for notifying `PassManager` of any newly added loops so that it can put those new loops into the queue before processing them later. It can also tell the PassManager to put the current loop into the queue again.

When we are writing a pass, we will want to use the analysis data provided by AnalysisManager – in this case, it is the LoopAnalysisManager instance. LoopAnalysisManager has a similar usage to other versions of AnalysisManager (FunctionAnalysisManager, for example) we learned about in the previous chapter. The only difference is that we need to supply an additional argument to the getResult method. Here is an example:

```
PreservedAnalyses run(Loop &LP, LoopAnalysisManager &LAM,
                      LoopStandardAnalysisResults &LAR,
                      LPMUpdater &U) {
  ...
  LoopNest &LN = LAM.getResult<LoopNestAnalysis>(LP, LAR);
  ...
}
```

LoopNest is the analysis data that's generated by LoopNestAnalysis. (We will talk about both shortly in the *Dealing with nested loops* section.)

As shown in the previous snippet, LoopAnalysisManager::getResult takes another LoopStandarAnalysisResults type argument, in addition to the Loop instance.

Except for having different a run method signature and a slightly different usage of LoopAnalysisManager, developers can build their loop passes in the same way as other kinds of passes. Now that we've looked at the foundation provided by loop pass and AnalysisManager, it's time to look at some specialized loops. The first one we are going to introduce is the *nested* loop.

Dealing with nested loops

So far, we have been talking about loops with only one layer. However, nested loops – loops with other loop(s) enclosed in them – are also common in real-world scenarios. For example, most of the matrix multiplication implementations require at least two layers of loops.

Nested loops are usually depicted as a tree – called a **loop tree**. In a loop tree, every node represents a loop. If a node has a parent node, this means that the corresponding loop is *enclosed* within the loop being modeled by the parent. The following diagram shows an example of this:

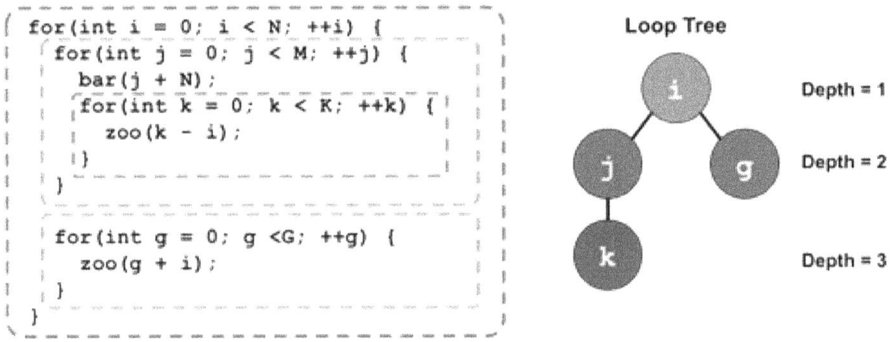

Figure 10.10 – A loop tree example

In the preceding diagram, loops j and g are enclosed within loop i, so they are both child nodes of loop i in the loop tree. Similarly, loop k – the innermost loop – is modeled as the child node of loop j in the tree.

The root of a loop tree also represents a *top-level* loop in a Function. Recall that, previously, we learned how to retrieve all Loop instances in a Function by iterating through the LoopInfo object – each of the Loop instances that were retrieved in this way are top-level loops. For a given Loop instance, we can retrieve its subloops at the next layer in a similar way. Here is an example:

```
// `LP` has the type of `Loop&`
for (Loop *SubLP : LP) {
    // `SubLP` is one of the sub-loops at the next layer
}
```

Note that the preceding snippet only traversed the subloops at the next level, rather than all the descendant subloops. To traverse all descendant subloops in a tree, you have two options:

- By using the `Loop::getLoopsInPreorder()` method, you can traverse all the descendant loops of a `Loop` instance in a pre-ordered fashion.

- In the *Iterating different IR units* section, we have learned what `GraphTraits` is and how LLVM uses it for graph traversal. It turns out that LLVM also has a default implementation of `GraphTraits` for the loop tree. Therefore, you can traverse a loop tree with existing graph iterators in LLVM, such as post-ordering and depth-first, to name a few. For example, the following code tries to traverse a loop tree rooted at `RootL` in a depth-first fashion:

```
#include "llvm/Analysis/LoopInfo.h"
#include "llvm/ADT/DepthFirstIterator.h"
...
// `RootL` has the type of `Loop*`
for (Loop *L : depth_first(RootL)) {
  // Work with `L`
}
```

With the help of `GraphTraits`, we can have more flexibility when it comes to traversing a loop tree.

In addition to dealing with individual loops in a loop tree, LLVM also provides a wrapper class that represents the whole structure – `LoopNest`.

`LoopNest` is analysis data that's generated by `LoopNestAnalysis`. It encapsulates all the subloops in a given `Loop` instance and provides several "shortcut" APIs for commonly used functionalities. Here are some of the important ones:

- `getOutermostLoop()`/`getInnermostLoop()`: These utilities retrieve the outer/innermost `Loop` instances. These are pretty handy because many loop optimizations only apply to either the inner or outermost loop.

- `areAllLoopsSimplifyForm()`/`areAllLoopsRotatedForm()`: These useful utilities tell you if all the enclosing loops are in a certain canonical form, as we mentioned in the previous section.

- `getPerfectLoops (...)`: You can use this to get all the *perfect loops* in the current loop hierarchy. By perfect loops, we are referring to loops that are nested together without a "gap" between them. Here is an example of perfect loops and non-perfect loops:

```
// Perfect loops
for(int i=…) {
   for(int j=…){…}
}
// Non-perfect loops
for(int x=…) {
   foo();
   for(int y=…){…}
}
```

In the non-perfect loops example, the `foo` call site is the gap between the upper and lower loops.

Perfect loops are preferrable in many loop optimizations. For example, it's easier to *unroll* perfectly nested loops – ideally, we only need to duplicate the body of the innermost loop.

With that, you've learned how to work with nested loops. In the next section, we are going to learn about another important topic for loop optimization: induction variables.

Retrieving induction variables and their range

The induction variable is a variable that progresses by a certain pattern in each loop iteration. It is the key to many loop optimizations. For example, in order to *vectorize* a loop, we need to know how the induction variable is used by the array – the data we want to put in a vector – within the loop. The induction variable can also help us resolve the **trip count** – the total number of iterations – of a loop. Before we dive into the details, the following diagram shows some terminology related to induction variables and where they're located in the loop:

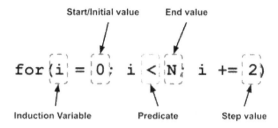

Figure 10.11 – Terminology for an induction variable

Now, let's introduce some APIs that can help you retrieve the components shown in the preceding diagram.

First, let's talk about the induction variable. The Loop class already provides two convenient methods for retrieving the induction variable: getCanonicalInductionVariable and getInductionVariable. Both methods return a PHINode instance as the induction variable (if there are any). The first method can only be used if the induction variable starts from zero and only increments by one on each iteration. On the other hand, the second method can handle more complicated cases, but requires a ScalarEvolution instance as the argument.

ScalarEvolution is an interesting and powerful framework in LLVM. Simply put, it tries to track how values *change* – for example, through arithmetic operation – over the program path. Putting this into the context of loop optimization, it is used to capture recurrence value-changing behaviors in the loop, which has a strong relationship with the induction variable.

To find out more about the induction variable's behavior in a loop, you can retrieve an InductionDescriptor instance via Loop::getInductionDescriptor. An InductionDescriptor instance provides information such as the initial value, the step value, and the instruction that *updates* the induction variable at each iteration. The Loop class also provides another similar data structure for realizing the boundaries of the induction variable: the Loop::LoopBounds class. LoopBounds not only provides the initial and step values of the induction variable, but also the prospective ending value, as well as the predicate that's used for checking the exit condition. You can retrieve a LoopBounds instance via the Loop::getBounds method.

Loops are crucial for a program's runtime performance. In this section, we learned how loops are represented in LLVM IR and how to work with them. We also looked at their high-level concepts and various practical APIs for retrieving the desired loop properties. With this knowledge, you are one step closer to creating a more effective, aggressive loop optimization and gaining even higher performance from your target applications.

Summary

In this section, we learned about LLVM IR – the target-independent intermediate representation that sits at the core of the entire LLVM framework. We provided an introduction to the high-level structure of LLVM IR, followed by practical guidelines on how to walk through different units within its hierarchy. We also focused on instructions, values, and SSA form at, which are crucial for working with LLVM IR efficiently. We also presented several practical skills, tips, and examples on the same topic. Last but not least, we learned how to process loops in LLVM IR – an important technique for optimizing performance-sensitive applications. With these abilities, you can perform a wider range of program analysis and code optimization tasks on LLVM IR.

In the next chapter, we will learn about a collection of LLVM utilities APIs that can improve your productivity when it comes to developing, diagnosing, and debugging with LLVM.

11
Gearing Up with Support Utilities

In the previous chapter, we learned the basics of **Low-Level Virtual Machine** (**LLVM**) **intermediate representation** (**IR**)—the target-independent intermediate representations in LLVM—and how to inspect and manipulate this with C++ **application programming interfaces** (**APIs**). These are the core techniques for doing program analysis and transformation in LLVM. In addition to those skill sets, LLVM also provides many support utilities to improve compiler developers' productivity when working with LLVM IR. We are going to cover those topics in this chapter.

A compiler is a complex piece of software. It not only needs to handle thousands of different cases— including input programs with different shapes and a wide variety of target architectures—but the **correctness** of a compiler is also an important topic: namely, the compiled code needs to have the same behavior as the original one. LLVM, a large-scale compiler framework (and probably one of the biggest), is not an exception.

To tackle these complexities, LLVM has provided a crate of gadgets to improve the development experience. In this chapter, we are going to show you how to gear up to use those tools. The utilities covered here can assist you in diagnosing problems that occur from the LLVM code you are developing. This includes more efficient debugging, error handling, and profiling abilities; for instance, one of the tools can collect statistical numbers on key components—such as the number of basic blocks being processed by a specific Pass—and automatically generate a summary report. Another example is LLVM's own error-handling framework, which prevents as many unhandled errors (a common programming mistake) as possible.

Here is a list of the topics we are going to cover in this chapter:

- Printing diagnostic messages
- Collecting statistics
- Adding time measurements
- Error-handling utilities in LLVM
- Learning about the `Expected` and `ErrorOr` classes

With the help of these utilities, you will have a better time debugging and diagnosing the LLVM code, letting you focus on the core logic you want to implement with LLVM.

Technical requirements

In this section, we are also going to use LLVM Pass as the platform to show different API usages. Therefore, please make sure you have built the `opt` command-line tool, as follows:

```
$ ninja opt
```

Note that some of the content in this chapter only works with a **debug build** version of LLVM. Please check the first chapter, *Chapter 1, Saving Resources When Building LLVM,* for a recap on how to build LLVM in debug mode.

You can also go back to *Chapter 9, Working with PassManager and AnalysisManager,* if you are not sure how to create a new LLVM Pass.

The sample code for this chapter can be found here:

```
https://github.com/PacktPublishing/LLVM-Techniques-Tips-and-
Best-Practices-Clang-and-Middle-End-Libraries/tree/main/
Chapter11
```

Printing diagnostic messages

In software development, there are many ways to diagnose a bug—for instance, using a debugger, inserting a sanitizer into your program (to catch invalid memory access, for example), or simply using one of the simplest yet most effective ways: adding *print statements*. While the last option doesn't sound really smart, it is actually pretty useful in many cases where other options cannot unleash their full potential (for example, release mode binaries with poor debug information quality or multithread programs).

LLVM provides a small utility that not only helps you to print out debug messages but also *filters* which messages to show. Let's say we have an LLVM Pass, `SimpleMulOpt`, which replaces multiplication by power-of-two constants with left-shifting operations (which is what we did in the last section of the previous chapter, *Processing LLVM IR*). Here is part of its `run` method:

```
PreservedAnalyses
SimpleMulOpt::run(Function &F, FunctionAnalysisManager &FAM) {
  for (auto &I : instructions(F)) {
    if (auto *BinOp = dyn_cast<BinaryOperator>(&I) &&
        BinOp->getOpcode() == Instruction::Mul) {
      auto *LHS = BinOp->getOperand(0),
           *RHS = BinOp->getOperand(1);
      // `BinOp` is a multiplication, `LHS` and `RHS` are its
      // operands, now trying to optimize this instruction…
      …
    }
  }
  …
}
```

The preceding code iterates through all instructions in the given function before looking for instructions that represent arithmetic multiplication. If there are any, the Pass will then work with the `LHS` and `RHS` operands (which appear in the rest of the code—these are not shown here).

Let's assume that we want to print out the operand variables during our development. The most naïve way will be by using our old friend `errs()`, which streams arbitrary messages to `stderr`, as shown in the following code snippet:

```
// (extracted from the previous snippet)
…
auto *LHS = BinOp->getOperand(0),
     *RHS = BinOp->getOperand(1);
errs() << "Found a multiplication with operands ";
LHS->printAsOperand(errs());
errs() << " and ";
RHS->printAsOperand(errs());
…
```

The `printAsOperand` used in the preceding code snippet prints the textual representation of a `Value` to the given stream (`errs()`, in this case).

Everything looks normal, except the fact that these messages will be printed out anyway even in a production environment, which is not what we want. Either we need to remove these codes before we ship our products, adding some macro guard around these codes (for example, `#ifndef NDEBUG`), or we can use the debug utility provided by LLVM. Here is an example of this:

```
#include "llvm/Support/Debug.h"
#define DEBUG_TYPE "simple-mul-opt"
…
auto *LHS = BinOp->getOperand(0),
     *RHS = BinOp->getOperand(1);
LLVM_DEBUG(dbgs() << "Found a multiplication with operands ");
LLVM_DEBUG(LHS->printAsOperand(dbgs()));
LLVM_DEBUG(dbgs() << " and ");
LLVM_DEBUG(RHS->printAsOperand(dbgs()));
…
```

The preceding code is basically doing the following three things:

- Replacing any usage of `errs()` with `dbgs()`. These two streams are basically doing the same thing, but the latter one will add a nice banner (`Debug Log Output`) to the output message.

- Wrapping all lines related to debug printing with the `LLVM_DEBUG(...)` macro function. The use of this macro ensures that the enclosing line is only compiled in development mode. It also encodes the debug message category, which we will introduce shortly.

- Before using any `LLVM_DEBUG(...)` macro functions, please make sure you define `DEBUG_TYPE` to the desired debug category string (`simple-mul-opt`, in this case).

In addition to the aforementioned code modification, we also need to use an additional command-line flag, `-debug`, with `opt` to print those debug messages. Here is an example of this:

```
$ opt -O3 -debug -load-pass-plugin=… …
```

But then, you'll find the output to be pretty noisy. There are tons of debug messages from *other* LLVM Passes. In this case, we're only interested in the messages from our Pass.

To filter out unrelated messages, we can use the `-debug-only` command-line flag. Here is an example of this:

```
$ opt -O3 -debug-only=simple-mul-opt -load-pass-plugin=… …
```

The value after `-debug-only` is the `DEBUG_TYPE` value we defined in the previous code snippet. In other words, we can use `DEBUG_TYPE` defined by each Pass to filter the desired debug messages. We can also select *multiple* debug categories to print. For instance, check out the following command:

```
$ opt -O3 -debug-only=sroa,simple-mul-opt -load-pass-plugin=… …
```

This command not only prints debug messages from our `SimpleMulOpt` Pass, but also those coming from the `SROA` Pass—an LLVM Pass included in the `O3` optimization pipeline.

In addition to defining a single debug category (DEBUG_TYPE) for an LLVM Pass, you are in fact free to use as many categories as you like inside a Pass. This is useful, for instance, when you want to use separate debug categories for different parts of a Pass. For example, we can use separate categories for each of the operands in our SimpleMulOpt Pass. Here is how we can do this:

```
...
#define DEBUG_TYPE "simple-mul-opt"
auto *LHS = BinOp->getOperand(0),
     *RHS = BinOp->getOperand(1);
LLVM_DEBUG(dbgs() << "Found a multiplication instruction");
DEBUG_WITH_TYPE("simple-mul-opt-lhs",
                LHS->printAsOperand(dbgs() << "LHS operand: "));
DEBUG_WITH_TYPE("simple-mul-opt-rhs",
                RHS->printAsOperand(dbgs() << "RHS operand: "));
...
```

DEBUG_WITH_TYPE is a special version of LLVM_DEBUG. It executes code at the second argument, with the first argument as the debug category, which can be different from the currently defined DEBUG_TYPE value. In the preceding code snippet, in addition to printing Found a multiplication instruction using the original simple-mul-opt category, we are using simple-mul-opt-lhs to print messages related to the **left-hand-side** (**LHS**) operand and use simple-mul-opt-rhs to print messages for the other operand. With this feature, we can have a finer granularity to select debug message categories via the opt command.

You have now learned how to use the utility provided by LLVM to print out debug messages in the development environment only, and how to filter them if needed. In the next section, we are going to learn how to collect key statistics while running an LLVM Pass.

Collecting statistics

As mentioned in the previous section, a compiler is a complex piece of software. Collecting **statistical numbers**—for example, the number of basic blocks processed by a specific optimization—is one of the easiest and most efficient ways to get a quick portrait on the runtime behaviors of a compiler.

There are several ways to collect statistics in LLVM. In this section, we are going to learn three of the most common and useful options for doing this, and these methods are outlined here:

- Using the `Statistic` class
- Using an optimization remark
- Adding time measurements

The first option is a general utility that collects statistics via simple counters; the second option is specifically designed to profile *compiler optimizations*; and the last option is used for collecting timing information in the compiler.

Let's start with the first one.

Using the Statistic class

In this section, we are going to demonstrate new features by amending them to the `SimpleMulOpt` LLVM Pass from the previous section. First, let's assume that we don't only want to print out the operand `Value` from multiplication instructions but that we also want to *count* how many multiplication instructions have been processed by our Pass. First, let's try to implement this feature using the `LLVM_DEBUG` infrastructure we just learned about, as follows:

```
#define DEBUG_TYPE "simple-mul-opt"
PreservedAnalyses
SimpleMulOpt::run(Function &F, FunctionAnalysisManager &FAM) {
  unsigned NumMul = 0;
  for (auto &I : instructions(F)) {
    if (auto *BinOp = dyn_cast<BinaryOperator>(&I) &&
        BinOp->getOpcode() == Instruction::Mul) {
      ++NumMul;
      ...
    }
  }
  LLVM_DEBUG(dbgs() << "Number of multiplication: " << NumMul);
  ...
}
```

This approach seems pretty straightforward. But it comes with a drawback—the statistical numbers we are interested in are mixed with other debug messages. We need to take additional actions to parse or filter the value we want because although you might argue that these problems could be tackled by using a separate DEBUG_TYPE tag for each counter variable, when the number of counter variables increases, you might find yourself creating lots of redundant code.

One elegant solution is to use the Statistic class (and related utilities) provided by LLVM. Here is a version rewritten using this solution:

```cpp
#include "llvm/ADT/Statistic.h"
#define DEBUG_TYPE "simple-mul-opt"
STATISTIC(NumMul, "Number of multiplications processed");
PreservedAnalyses
SimpleMulOpt::run(Function &F, FunctionAnalysisManager &FAM) {
  for (auto &I : instructions(F)) {
    if (auto *BinOp = dyn_cast<BinaryOperator>(&I) &&
        BinOp->getOpcode() == Instruction::Mul) {
      ++NumMul;
      ...
    }
  }
  ...
}
```

The preceding code snippet shows the usage of Statistic, calling the STATISTIC macro function to create a Statistic type variable (with a textual description) and simply using it like a normal integer counter variable.

This solution only needs to modify a few lines in the original code, plus it collects all counter values and prints them in a table view at the end of the optimization. For example, if you run the SimpleMulOpt Pass using the -stats flag with opt, you will get the following output:

```
$ opt -stats -load-pass-plugin=… …
===-------------------------------===
       … Statistics Collected …
===-------------------------------===
87 simple-mul-opt - Number of multiplications processed
$
```

`87` is the number of multiplication instructions processed in `SimpleMulOpt`. Of course, you are free to add as many `Statistic` counters as you want in order to collect different statistics. If you run more than one Pass in the pipeline, all of the statistical numbers will be presented in the same table. For instance, if we add another `Statistic` counter into `SimpleMulOpt` to collect a number of `none-power-of-two constant operands` from the multiplication instructions and run the Pass with **Scalar Replacement of Aggregates (SROA)**, we can get an output similar to the one shown next:

```
$ opt -stats –load-pass-plugin=… --passes="sroa,simple-mult-
opt" …

===---------------------------------===
      … Statistics Collected …
===---------------------------------===
94   simple-mul-opt - Number of multiplications processed
87   simple-mul-opt - Number of none-power-of-two constant
operands
100 sroa            - Number of alloca partition uses rewritten
34  sroa            - Number of instructions deleted
...
$
```

The second column in the preceding code snippet is the name of the origin Pass, which is designated by the `DEBUG_TYPE` value defined prior to any calls to `STATISTIC`.

Alternatively, you can output the result in **JavaScript Object Notation (JSON)** format by adding the `-stats-json` flag to `opt`. For example, look at the following code snippet:

```
$ opt -stats -stats-json –load-pass-plugin=… …
{
        "simple-mul-opt.NumMul": 87
}
$
```

In this JSON format, instead of printing statistic values with a textual description, the field name of a statistic entry has this format: `"<Pass name>.<Statistic variable name>"` (the Pass name here is also the value of `DEBUG_TYPE`). Furthermore, you can print statistic results (either in default or JSON format) into a file using the `-info-output-file=<file name>` command-line option. The following code snippet shows an example of this:

```
$ opt -stats -stats-json -info-output-file=my_stats.json …
$ cat my_stats.json
{
        "simple-mul-opt.NumMul": 87
}
$
```

You have now learned how to collect simple statistic values using the `Statistic` class. In the next section, we are going to learn a statistic collecting method that is unique to compiler optimization.

Using an optimization remark

A typical compiler optimization usually consists of two stages: *searching* for the desired patterns from the input code, followed by *modifying* the code. Take our `SimpleMulOpt` Pass as an example: the first stage is to look for multiplication instructions (`BinaryOperator` with the `Instruction::Mul` **operation code (opcode)**) with power-of-two constant operands. For the second stage, we create new left-shifting instructions via `IRBuilder::CreateShl(…)` and replace all old usages of multiplication instructions with these.

There are many cases, however, where the optimization algorithm simply "bails out" during the first stage due to *infeasible* input code. For example, in `SimpleMulOpt`, we are looking for a multiplication instruction, but if the incoming instruction is not `BinaryOperator`, the Pass will not proceed to the second stage (and continue on to the next instruction). Sometimes, we want to know the *reason* behind this bailout, which can help us to improve the optimization algorithm or diagnose incorrect/suboptimal compiler optimization. LLVM provides a nice utility called an **optimization remarks** to collect and report this kind of bailout (or any kind of information) occurring in optimization Passes.

For example, let's assume we have the following input code:

```
int foo(int *a, int N) {
  int x = a[5];
  for (int i = 0; i < N; i += 3) {
```

```
    a[i] += 2;
    x = a[5];
  }
  return x;
}
```

Theoretically, we can use **loop-invariant code motion** (**LICM**) to optimize this code into an equivalent code base such as this one:

```
int foo(int *a, int N) {
  for (int i = 0; i < N; i += 3) {
    a[i] += 2;
  }
  return a[5];
}
```

We can do this as the fifth array element, a[5], never changed its value inside the loop. However, if we run LLVM's LICM Pass over the original code, it fails to perform the expected optimization.

To diagnose this problem, we can invoke the opt command with an additional option: --pass-remarks-output=<filename>. The filename will be a **YAML Ain't Markup Language** (**YAML**) file in which optimization remarks print out the possible reasons why LICM failed to optimize. Here is an example of this:

```
$ opt -licm input.ll -pass-remarks-output=licm_remarks.yaml …
$ cat licm_remarks.yaml
…
--- !Missed
Pass:            licm
Name:            LoadWithLoopInvariantAddressInvalidated
Function:        foo
Args:
  - String:          failed to move load with loop-invariant
address because the loop may invalidate its value
...
$
```

The `cat` command in the preceding output shows one of the optimization remark entries in `licm_remarks.yaml`. This entry tells us that there was a *missed* optimization that happened in the LICM Pass when it was processing the `foo` function. It also tells us the reason: LICM was not sure if a particular memory address was invalidated by the loop. Though this message doesn't provide fine-grained details, we can still infer that the problematic memory address concerning LICM was probably a `[5]`. LICM was not sure if the `a[i] += 2` statement modified the content of `a[5]`.

With this knowledge, compiler developers can get hands-on in improving LICM—for example, teaching LICM to recognize induction variables (that is, the `i` variable in this loop) with a step value greater than 1 (in this case, it was 3, since `i += 3`).

To generate optimization remarks such as the one shown in the preceding output, compiler developers need to integrate a specific utility API into their optimization Pass. To show you how to do that in your own Pass, we are going to reuse our `SimpleMulOpt` Pass as the sample. Here is part of the code that performs the first stage—*searching for multiplications* with power-of-two constant operands—in `SimpleMulOpt`:

```
...
for (auto &I : instructions(F)) {
  if (auto *BinOp = dyn_cast<BinaryOperator>(&I))
    if (BinOp->getOpcode() == Instruction::Mul) {
      auto *LHS = BinOp->getOperand(0),
           *RHS = BinOp->getOperand(1);
      // Has no constant operand
      if (!isa<Constant>(RHS)) continue;
      const APInt &Const = cast<ConstantInt>(RHS)->getValue();
      // Constant operand is not power of two
      if (!Const.isPowerOf2()) continue;
      ...
    }
}
```

The preceding code checks if the operand is constant before making sure it's also a power-of-two operand. If either of these checks fails, the algorithm will bail out by continuing on to the next instruction in the function.

We intentionally inserted a small flaw into this code to make it less powerful, and we are going to show you how to find that problem by using an optimization remark. Here are the steps to do this:

1. First, we need to have an `OptimizationRemarkEmitter` instance, which can help you to emit remark messages. This can be obtained from its parent analyzer, `OptimizationRemarkEmitterAnalysis`. Here is how we include it at the beginning of the `SimpleMulOpt::run` method:

```cpp
#include "llvm/Analysis/OptimizationRemarkEmitter.h"
PreservedAnalyses
SimpleMulOpt::run(Function &F, FunctionAnalysisManager
&FAM) {
  OptimizationRemarkEmitter &ORE
    = FAM.getResult<OptimizationRemarkEmitterAnalysis>(F);
  …
}
```

2. Then, we are going to use this `OptimizationRemarkEmitter` instance to emit an optimization remark if the multiplication instruction lacks a constant operand, as follows:

```cpp
#include "llvm/IR/DiagnosticInfo.h"
…
if (auto *BinOp = dyn_cast<BinaryOperator>(&I))
  if (BinOp->getOpcode() == Instruction::Mul) {
    auto *LHS = BinOp->getOperand(0),
         *RHS = BinOp->getOperand(1);
    // Has no constant operand
    if (!isa<ConstantInt>(RHS)) {
      std::string InstStr;
      raw_string_ostream SS(InstStr);
      I.print(SS);
      ORE.emit([&]() {
        return OptimizationRemarkMissed(DEBUG_TYPE,
                                        "NoConstOperand", &F)
          << "Instruction" <<
          << ore::NV("Inst", SS.str())
          << " does not have any constant operand";
```

```
        });
        continue;
    }
  }
  …
```

There are several things to be noticed here, as follows:

- The `OptimizationRemarkEmitter::emit` method takes a lambda function as the argument. This lambda function will be invoked to emit an optimization remark object if the optimization remark feature is turned on (via the –`pass-remarks-output` command-line option we've seen previously, for example).

- The `OptimizationRemarkMissed` class (note that it is not declared in `OptimizationRemarkEmitter.h` but in the `DiagnosticInfo.h` header file) represents the remark of a missed **optimization opportunity**. In this case, the missed opportunity is the fact that instruction `I` does not have any constant operand. The constructor of `OptimizationRemarkMissed` takes three arguments: the name of the Pass, the name of the missed optimization opportunity, and the enclosing IR unit (in this case, we use the enclosing `Function`). In addition to constructing a `OptimizationRemarkMissed` object, we also concatenate several objects via the stream operator (`<<`) at the tail. These objects will eventually be put under the `Args` section of each optimization remark entry in the YAML file we saw previously.

In addition to using `OptimizationRemarkMissed` to notify you of missed optimization opportunities, you can also use other classes derived from `DiagnosticInfoOptimizationBase` to present different kinds of information—for example, use `OptimizationRemark` to find out which optimization has been *successfully* applied, and use `OptimizationRemarkAnalysis` to keep a log of analysis data/facts.

- Among objects concatenated by the stream operator, `ore::NV(…)` seems to be a special case. Recall that in the optimization remark YAML file, each line under the `Args` section was a key-value pair (for example, `String:  failed to move load with…`, where `String` was the key). The `ore::NV` object allows you to customize the key-value pair. In this case, we are using `Inst` as the key and `SS.str()` as the value. This feature provides more flexibility to parse the optimization remark YAML file—for instance, if you want to write a little tool to visualize the optimization remarks, custom `Args` keys can give you an easier time (during the parsing stage) by distinguishing critical data from other strings.

3. Now that you have inserted the code to emit the optimization remark, it's time to test it. This time, we are going to use the following `IR` function as the input code:

```
define i32 @bar(i32 %0) {
    %2 = mul nsw i32 %0, 3
    %3 = mul nsw i32 8, %3
    ret %3
}
```

You can rebuild the `SimpleMulOpt` Pass and run it using a command such as this:

```
$ opt -load-pass-plugin=… -passes="simple-mul-opt" \
      -pass-remarks-output=remark.yaml -disable-output
      input.ll
$ cat remark.yaml
--- !Missed
Pass:           simple-mul-opt
Name:           NoConstOperand
Function:       bar
Args:
  - String:         'Instruction'
  - Inst:           ' %3 = mul nsw i32 8, %3'
  - String:         ' does not contain any constant
operand'
...
$
```

From this optimization remark entry, we can glean that `SimpleMulOpt` bailed out because it couldn't find a constant operand on one of the (multiplication) instructions. The `Args` section shows a detailed reason for this.

With this information, we realize that `SimpleMulOpt` is unable to optimize a multiplication whose *first* operand (LHS operand) is a power-of-two constant, albeit a proper optimization opportunity. Thus, we can now fix the implementation of `SimpleMulOpt` to check if *either* of the operands is constant, as follows:

```
…
if (BinOp->getOpcode() == Instruction::Mul) {
    auto *LHS = BinOp->getOperand(0),
         *RHS = BinOp->getOperand(1);
    // Has no constant operand
```

```
if (!isa<ConstantInt>(RHS) && !isa<ConstantInt>(LHS)) {
  ORE.emit([&]() {
    return …
  });
  continue;
}
…
}
…
```

You have now learned how to emit optimization remarks in an LLVM Pass and how to use the generated report to discover potential optimization opportunities.

So far, we have only studied the generated optimization remark YAML file. Though it has provided valuable diagnostic information, it would be great if we could have more fine-grained and intuitive location information to know where exactly these remarks happened. Luckily, Clang and LLVM have provided a way to achieve that.

With the help of Clang, we can actually generate optimization remarks with **source location** (that is, line and column numbers in the original source file) attached. Furthermore, LLVM provides you with a small utility that can associate an optimization remark with its corresponding source location and visualize the result on a web page. Here's how to do this:

1. Let's reuse the following code as the input:

```
int foo(int *a, int N) {
  for (int i = 0; i < N; i += 3) {
    a[i] += 2;
  }
  return a[5];
}
```

First, let's generate optimization remarks using this `clang` command:

```
$ clang -O3 -foptimization-record-file=licm.remark.yaml \
     -S opt_remark_licm.c
```

Though we're using a different name, `-foptimization-record-file` is the command-line option used to generate an optimization remark file with the given filename.

2. After `licm.remark.yaml` is generated, let's use a utility called `opt-viewer.py` to visualize the remarks. The `opt-viewer.py` script is not installed in the typical location by default—instead of putting it in `<install path>/bin` (for example `/usr/bin`), it is installed in `<install path>/share/opt-viewer` (`/usr/share/opt-viewer`). We are going to invoke this script with the following command-line options:

```
$ opt-viewer.py --source-dir=$PWD \
--target-dir=licm_remark licm.remark.yaml
```

(Note that `opt-viewer.py` depends on several Python packages such as `pyyaml` and `pygments`. Please install them before you use `opt-viewer.py`.)

3. There will be a HTML file—`index.html`—generated inside the `licm_remark` folder. Before you open the web page, please copy the original source code—`opt_remark_licm.c`—into that folder as well. After that, you will be able to see a web page like this:

Source Location	Hotness	Function	Pass
./opt_remark_licm.c:0:0		foo	asm-printer
./opt_remark_licm.c:0:0		foo	gvn
./opt_remark_licm.c:1:0		foo	asm-printer
./opt_remark_licm.c:1:0		foo	prologepilog
./opt_remark_licm.c:3:3		foo	asm-printer
./opt_remark_licm.c:3:3		foo	asm-printer
./opt_remark_licm.c:3:3		foo	asm-printer
./opt_remark_licm.c:3:3		foo	asm-printer
./opt_remark_licm.c:3:3		foo	loop-unroll
./opt_remark_licm.c:3:3		foo	loop-vectorize
./opt_remark_licm.c:3:3		foo	loop-vectorize
./opt_remark_licm.c:3:21		foo	asm-printer
./opt_remark_licm.c:4:5		foo	licm
./opt_remark_licm.c:4:5		foo	licm
./opt_remark_licm.c:4:10		foo	asm-printer
./opt_remark_licm.c:4:10		foo	asm-printer

Figure 11.1 – Web page of optimization remarks combined with the source file

We are particularly interested in two of these columns: **Source Location** and **Pass**. The latter column shows the name of the Pass and the type of the optimization remark—`Missed`, `Passed`, or `Analyzed` rendered in red, green, and white, respectively—attached on a given line shown at the **Source Location** column.

If we click on a link in the **Source Location** column, this will navigate you to a page that looks like this:

LineHotness	Optimization	Source	Inline Context
1		int foo(int *a, int N) {	
	prologepilog	0 stack bytes in function	foo
	asm-printer	37 instructions in function	foo
2		int x = a[5];	
3		for (int i = 0; i < N; i += 3) {	
	loop-vectorize	the cost-model indicates that vectorization is not beneficial	foo
	loop-vectorize	the cost-model indicates that interleaving is not beneficial	foo
	loop-unroll	unrolled loop by a factor of 4 with run-time trip count	foo
	asm-printer	+ BasicBlock:	foo
	asm-printer	+ BasicBlock:	foo
	asm-printer	+ BasicBlock:	foo
	asm-printer	+ BasicBlock:	foo
	asm-printer	+ BasicBlock:	foo
4		a[i] += 2;	
	licm	sinking getelementptr	foo
	licm	sinking zext	foo
	asm-printer	+ BasicBlock:	foo
	asm-printer	+ BasicBlock:	foo
5		x = a[5];	
6		}	
7		return x;	
8		}	
9			
10			

Figure 11.2 – Details of an optimization remark

This page gives you a nice view of optimization remark details, interleaved with the originating source code line. For example, on *line 3*, `loop-vectorize` Pass said it couldn't vectorize this loop because its cost model didn't think it was beneficial to do so.

You have now learned how to use optimization remarks to gain insights into the optimization Pass, which is especially useful when you're debugging a missing optimization opportunity or fixing a mis-compilation bug.

In the next section, we are going to learn some useful skills to profile the execution time of LLVM.

Adding time measurements

LLVM is an enormous software, with hundreds of components working closely together. Its ever-increasing running time is slowly becoming an issue. This affects many use cases that are sensitive to compilation time—for example, the **Just-in-Time (JIT)** compiler. To diagnose this problem in a systematic way, LLVM provides some useful utilities for **profiling** the execution time.

Time profiling has always been an important topic in software development. With the running time collected from individual software components, we can spot performance bottlenecks more easily. In this section, we are going to learn about two tools provided by LLVM: the `Timer` class and the `TimeTraceScope` class. Let's start with the `Timer` class first.

Using the Timer class

The `Timer` class, as suggested by its name, can measure the execution time of a code region. Here is an example of this:

```
#include "llvm/Support/Timer.h"
...
Timer T("MyTimer", "A simple timer");
T.startTimer();
// Do some time-consuming works…
T.stopTimer();
```

In the preceding snippet, `Timer` instance `T` measures the time spent in the region, enclosed by the `startTimer` and `stopTimer` method calls.

Now that we have collected the timing data, let's try to print it out. Here is an example of this:

```
Timer T(…);
...
TimeRecord TR = T.getTotalTime();
TR.print(TR, errs());
```

In the previous code snippet, a `TimeRecord` instance encapsulates the data collected by the `Timer` class. We can then use `TimeRecord::print` to print it to a stream—in this case, the `errs()` stream. In addition, we assigned another `TimeRecord` instance—via the first argument of `print`—as the *total* time interval we want to compare it against. Let's look at the output of this code, as follows:

```
===-------------------------------------------------------------===
                     Miscellaneous Ungrouped Timers
===-------------------------------------------------------------===


    ---User Time---   --User+System--   ---Wall Time---   ---
Name ---
```

```
   0.0002 (100.0%)    0.0002 (100.0%)    0.0002 (100.0%)  A
simple timer
   0.0002 (100.0%)    0.0002 (100.0%)    0.0002 (100.0%)  Total


   0.0002 (100.0%)    0.0002 (100.0%)    0.0002 (100.0%)
```

In the preceding output, the first row shows the `TimeRecord` instance collected from our previous `Timer` instance, whereas the second row shows the total time—the first argument of `TimeRecord::print`.

We now know how to print the timing data collected by a single `Timer` instance, but what about multiple timers? LLVM provides another support utility for the `Timer` class: the `TimerGroup` class. Here's an example usage of the `TimerGroup` class:

```
TimerGroup TG("MyTimerGroup", "My collection of timers");

Timer T("MyTimer", "A simple timer", TG);
T.startTimer();
// Do some time-consuming works…
T.stopTimer();

Timer T2("MyTimer2", "Yet another simple timer", TG);
T2.startTimer();
// Do some time-consuming works…
T2.stopTimer();

TG.print(errs());
```

In the preceding code snippet, we declare a `TimerGroup` instance, TG, and use it as the third constructor argument for each `Timer` instance we create. Finally, we print them using `TimerGroup::print`. Here is the output of this code:

```
===----------------------------------------------------------===
                    My collection of timers
===----------------------------------------------------------===
   Total Execution Time: 0.0004 seconds (0.0004 wall clock)

     ---User Time---    --User+System--    ---Wall Time---    ---
Name ---
   0.0002 ( 62.8%)    0.0002 ( 62.8%)    0.0002 ( 62.8%)  A
```

```
simple timer
   0.0001 ( 37.2%)    0.0001 ( 37.2%)    0.0001 ( 37.2%)   Yet
another simple timer
   0.0004 (100.0%)    0.0004 (100.0%)    0.0004 (100.0%)   Total
```

Each row in the output (except the last one) is the `TimeRecord` instance for each `Timer` instance in this group.

So far, we have been using `Timer::startTimer` and `Timer::stopTimer` to toggle the timer. To make measuring the time interval within a code block—namely, the region enclosed with curly brackets { }—easier without manually calling those two methods, LLVM provides another utility that automatically starts the timer upon entering a code block and turns it off when exiting. Let's see how to use the `TimeRegion` class with the following sample code:

```
TimerGroup TG("MyTimerGroup", "My collection of timers");
{
  Timer T("MyTimer", "A simple timer", TG);
  TimeRegion TR(T);
  // Do some time-consuming works…
}
{
  Timer T("MyTimer2", "Yet another simple timer", TG);
  TimeRegion TR(T);
  // Do some time-consuming works…
}

TG.print(errs());
```

As you can see in the preceding snippet, instead of calling `startTimer`/`stopTimer`, we put the to-be-measured code into a separate code block and use a `TimeRegion` variable to automatically toggle the timer. This code will print out the same content as the previous example. With the help of `TimeRegion`, we can have a more concise syntax and avoid any mistakes where we *forget* to turn off the timer.

You have now learned how to use `Timer` and its supporting utilities to measure the execution time of a certain code region. In the next section, we are going to learn a more advanced form of time measurement that captures the hierarchical structure of the program.

Collecting the time trace

In the previous section, we learned how to use `Timer` to collect the execution time of a small range of code regions. Although that gave us a portrait of the compiler's runtime performance, we sometimes need a more *structural* timing profile in order to fully understand any systematic issues.

`TimeTraceScope` is a class provided by LLVM to perform global-scope time profiling. Its usage is pretty simple: similar to `TimeRegion`, which we saw in the previous section, a `TimeTraceScope` instance automatically turns the time profiler on and off upon entering and exiting a code block. Here is an example of this:

```
TimeTraceScope OuterTimeScope("TheOuterScope");
for (int i = 0; i < 50; ++i) {
  {
    TimeTraceScope InnerTimeScope("TheInnerScope");
    foo();
  }
  bar();
}
```

In the preceding code snippet, we create two `TimeTraceScope` instances: `OuterTimeScope` and `InnerTimeScope`. These try to profile the execution time of the whole region and the time spent on function `foo`, respectively.

Normally, if we use `Timer` rather than `TimeTraceScope`, it can only give us the aggregate duration collected from each timer. However, in this case, we are more interested in how different parts of the code allocate themselves on the *timeline*. For example, does the `foo` function always spend the same amount of time ion every loop iteration? If that's not the case, which iterations spend more time than others?

To see the result, we need to add additional command-line options to the `opt` command when running the Pass (assuming you use `TimeTraceScope` within a Pass). Here is an example of this:

```
$ opt –passes="…" -time-trace -time-trace-file=my_trace.json …
```

The additional `-time-trace` flag is asking `opt` to export all the traces collected by `TimeTraceScope` to the file designated by the `-time-trace-file` option.

After running this command, you will get a new file, `my_trace.json`. The content of this file is basically non-human-readable, but guess what? You can visualize it using the **Chrome** web browser. Here are the steps to do this:

1. Open your Chrome web browser and type in `chrome://tracing` in the **Uniform Resource Locator** (**URL**) bar. You will see an interface that looks like this:

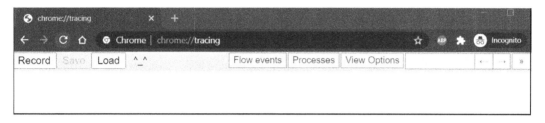

Figure 11.3 – The trace visualizer in Chrome

2. Click on the **Load** button in the top-left corner and select our `my_trace.json` file. You will see a page like this:

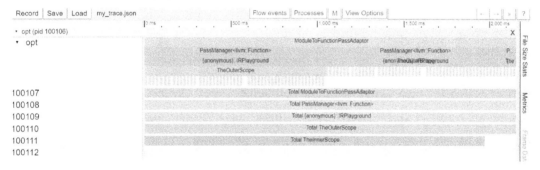

Figure 11.4 – The view after opening my_trace.json

Each color block represents a time interval collected by a `TimeTraceScope` instance.

3. Let's take a closer look: please press the number key *3* to switch to zoom mode. After that, you should be able to zoom in or out by clicking and dragging the mouse up or down. In the meantime, you can use the arrow keys to scroll the timeline left or right. Here is part of the timeline after we zoom in:

Figure 11.5 – Part of the trace timeline

As we can see from *Figure 11.5*, there are several layers stacking together. This layout reflects how different `TimeTraceScope` instances are organized in `opt` (and in our Pass). For example, our `TimeTraceScope` instance entitled `TheOuterScope` is stacked above multiple `TheInnerScope` blocks. Each of the `TheInnerScope` blocks represents the time spent on the `foo` function in each loop iteration we saw earlier.

4.   We can further inspect the properties of a block by clicking on it. For example, if we click one of the `TheInnerScope` blocks, its timing properties will be shown in the lower half of the screen. Here is an example of this:

1 item selected.	Slice (1)	
Title	TheInnerScope	🔍
User Friendly Category	other	
Start	948.107 ms	
Wall Duration	12.945 ms	

Figure 11.6 – Details of a time interval block

This gives us information such as the time interval and the starting time in this timeline.

With this visualization, we can combine timing information with the structure of the compiler, which will help us to find out performance bottlenecks more rapidly.

In addition to `opt`, `clang` can also generate the same trace JSON file. Please consider adding a `-ftime-trace` flag. Here is an example of this:

```
$ clang -O3 -ftime-trace -c foo.c
```

This will generate a JSON trace file with the same name as the input file. In this case, it will be `foo.json`. You can use the skills we just learned to visualize it.

In this section, we have learned some useful skills to collect statistics from LLVM. The `Statistic` class can be used as an integer counter to record the number of events occurring in the optimization. Optimization remarks, on the other hand, can give us insights into some of the decision-making process inside the optimization Pass, making it easier for compiler developers to diagnose missing optimization opportunities. With `Timer` and `TimeTraceScope`, developers can monitor LLVM's execution time in a more manageable way and handle compilation-speed regressions with confidence. These techniques can improve an LLVM developer's productivity when creating new inventions or fixing a challenging problem.

In the next section of this chapter, we are going to learn how to write error-handling code in an efficient way, using utilities provided by LLVM.

Error-handling utilities in LLVM

Error handling has always been a widely discussed topic in software development. It can be as simple as returning an error code—such as in many of the Linux APIs (for example, the open function)—or using an advanced mechanism such as throwing an exception, which has been widely adopted by many modern programming languages such as Java and C++.

Although C++ has built-in support for exception handling, LLVM does *not* adopt it in its code base at all. The rationale behind this decision is that despite its convenience and expressive syntax, exception handling in C++ comes at a high cost in terms of performance. Simply speaking, exception handling makes the original code more complicated and hinders a compiler's ability to optimize it. Furthermore, during runtime, the program usually needs to spend more time recovering from an exception. Therefore, LLVM disables exception handling by default in its code base and falls back to other ways of error handling—for example, carrying an error with the return value or using the utilities we are going to learn about in this section.

In the first half of this section, we are going to talk about the Error class, which—as the name suggests—represents an error. This is unlike conventional error representations—when using an integer as the error code, for instance, you cannot *ignore* the generated Error instances without handling it. We will explain this shortly.

In addition to the Error class, developers found that in LLVM's code base a common pattern was shared by much of the error-handling code: an API may return a result *or* an error, but not both (at the same time). For instance, when we call a file-reading API, we are expecting to get the content of that file (the result) or an error when something goes wrong (for example, there is no such file). In the second part of this section, we are going to learn two utility classes that implement this pattern.

Let's start with an introduction to the Error class first.

Introducing the Error class

The concept represented by the `Error` class is pretty simple: it's an error with supplementary descriptions such as an error message or error code. It is designed to be passed by a value (as a function argument) or returned from a function. Developers are free to create their custom `Error` instance, too. For example, if we want to create a `FileNotFoundError` instance to tell users that a certain file does not exist, we can write the following code:

```cpp
#include "llvm/Support/Error.h"
#include <system_error>
// In the header file...
struct FileNotFoundError : public ErrorInfo<FileNoteFoundError>
{
  StringRef FileName;
  explicit FileNotFoundError(StringRef Name) : FileName(Name)
    {}
  static char ID;
  std::error_code convertToErrorCode() const override {
    return std::errc::no_such_file_or_directory;
  }
  void log(raw_ostream &OS) const override {
    OS << FileName << ": No such file";
  }
};
// In the CPP file...
char FileNotFoundError::ID = 0;
```

There are several requirements for implementing a custom `Error` instance. These are listed next:

- Derive from the `ErrorInfo<T>` class, where `T` is your custom class.

- Declare a unique `ID` variable. In this case, we use a static class member variable.

- Implement the `convertToErrorCode` method. This method designates a `std::error_code` instance for this `Error` instance. `std::error_code` is the error type used in the C++ standard library (since C++11). Please refer to the C++ reference documentation for available (predefined) `std::error_code` instances.

- Implement the `log` method to print out error messages.

To create an `Error` instance, we can leverage a `make_error` utility function. Here is an example usage of this:

```
Error NoSuchFileErr = make_error<FileNotFoundError>("foo.txt");
```

The `make_error` function takes an error class—in this case, our `FileNotFoundError` class—as the template argument and function arguments (in this case, `foo.txt`), if there are any. These will then be passed to its constructor.

If you try to run the preceding code (in debug build) without doing anything to the `NoSuchFileErr` variable, the program will simply crash and show an error message such as this:

```
Program aborted due to an unhandled Error:
foo.txt: No such file
```

It turns out that every `Error` instance is required to be **checked** and **handled** before the end of its lifetime (that is, when its destructor method is called).

Let me first explain what *checking an* `Error` *instance* means. In addition to representing a real error, the `Error` class can also represent a *success* state—that is, no error. To give you a more concrete idea of this, many of the LLVM APIs have the following error-handling structure:

```
Error readFile(StringRef FileName) {
  if (openFile(FileName)) {
    // Success
    // Read the file content…
    return ErrorSuccess();
  } else
    return make_error<FileNotFoundError>(FileName);
}
```

In other words, they return an `ErrorSuccess` instance in the case of success or an `ErrorInfo` instance otherwise. When the program returns from `readFile`, we need to *check* if the returned `Error` instance represents a success result or not by treating it as a Boolean variable, as follows:

```
Error E = readFile(…);
if (E) {
  // TODO: Handle the error
} else {
```

```
  // Success!
}
```

Note that you *always* need to check an `Error` instance even if you are 100% sure that it is in a `Success` state, otherwise the program will still abort.

The preceding code snippet provides a good segue into the topic of handling `Error` instances. If an `Error` instance represents a real error, we need to use a special API to handle it: `handleErrors`. Here's how to use it:

```
Error E = readFile(…);
if (E) {
  Error UnhandledErr = handleErrors(
    std::move(E),
    [&](const FileNotFoundError &NotFound) {
      NotFound.log(errs() << "Error occurred: ");
      errs() << "\n";
    });
  …
}
```

The `handleErrors` function takes ownership of the `Error` instance (by `std::move(E)`) and uses the provided lambda function to handle the error. You might notice that `handleErrors` returns another `Error` instance, which represents the *unhandled* error. What does that mean?

In the previous example of the `readFile` function, the returned `Error` instance can represent either a `Success` state or a `FileNotFoundError` state. We can slightly modify the function to return a `FileEmptyError` instance when the opened file is empty, as follows:

```
Error readFile(StringRef FileName) {
  if (openFile(FileName)) {
    // Success
    …
    if (Buffer.empty())
      return make_error<FileEmptyError>();
    else
      return ErrorSuccess();
  } else
```

```
    return make_error<FileNotFoundError>(FileName);
}
```

Now, the `Error` instance returned from `readFile` can either be a `Success` state, a `FileNotFoundError` instance, *or* a `FileEmptyError` instance. However, the `handleErrors` code we wrote previously only handled the case of `FileNotFoundError`.

Therefore, we need to use the following code to handle the case of `FileEmptyError`:

```
Error E = readFile(…);
if (E) {
  Error UnhandledErr = handleErrors(
      std::move(E),
      [&](const FileNotFoundError &NotFound) {…});
  UnhandledErr = handleErrors(
      std::move(UnhandledErr),
      [&](const FileEmptyError &IsEmpty) {…});
  …
}
```

Be aware that you always need to take ownership of an `Error` instance when using `handleErrors`.

Alternatively, you can *coalesce* two `handleErrors` function calls into one by using multiple lambda function arguments for each of the error types, as follows:

```
Error E = readFile(…);
if (E) {
  Error UnhandledErr = handleErrors(
      std::move(E),
      [&](const FileNotFoundError &NotFound) {…},
      [&](const FileEmptyError &IsEmpty) {…});
  …
}
```

In other words, the handleErrors function is acting like a switch-case statement for an Error instance. It is effectively working like the following pseudocode:

```
Error E = readFile(…);
if (E) {
  switch (E) {
  case FileNotFoundError: …
  case FileEmptyError: …
  default:
    // generate the UnhandledError
  }
}
```

Now, you might be wondering: *Since* handleErrors *will always return an* Error *representing the unhandled error, and I can't just ignore the returned instance, otherwise the program will abort, how should we end this "chain of error handling"?* There are two ways to do that, so let's have a look at each, as follows:

- If you are 100% sure that you have handled all possible error types—which means that the unhandled Error variable is in a Success state—you can call the cantFail function to make an assertion, as illustrated in the following code snippet:

```
if (E) {
  Error UnhandledErr = handleErrors(
    std::move(E),
    [&](const FileNotFoundError &NotFound) {…},
    [&](const FileEmptyError &IsEmpty) {…});
  cantFail(UnhandledErr);
}
```

If UnhandledErr still contains an error, the cantFail function will abort the program execution and print an error message.

- A more elegant solution would be to use the handleAllErrors function, as follows:

```
if (E) {
  handleAllErrors(
    std::move(E),
    [&](const FileNotFoundError &NotFound) {…},
```

```
            [&](const FileEmptyError &IsEmpty) {…});
    …
    }
```

This function will not return anything. It assumes that the given lambda functions are sufficient to handle all possible error types. Of course, if there is a missing one, `handleAllErrors` will still abort the program execution, just like what we have seen previously.

You have now learned how to use the `Error` class and how to properly handle errors. Though the design of `Error` seems a little annoying at first glance (that is, we need to handle *all* possible error types or the execution will just abort halfway), these restrictions can decrease the number of mistakes made by programmers and create a more **robust** program.

Next, we are going to introduce two other utility classes that can further improve the error-handling expressions in LLVM.

Learning about the Expected and ErrorOr classes

As we briefly mentioned in the introduction of this section, in LLVM's code base it's pretty common to see a coding pattern where an API wants to return a result or an error if something goes wrong. LLVM tries to make this pattern more accessible by creating utilities that *multiplex* results and errors in a single object—they are the `Expected` and `ErrorOr` classes. Let's begin with the first one.

The Expected class

The `Expected` class carries either a `Success` result or an error—for instance, the JSON library in LLVM uses it to represent the outcome of parsing an incoming string, as shown next:

```
#include "llvm/Support/JSON.h"
  using namespace llvm;
  …
// `InputStr` has the type of `StringRef`
Expected<json::Value> JsonOrErr = json::parse(InputStr);
if (JsonOrErr) {
  // Success!
```

```
    json::Value &Json = *JsonOrErr;
    ...
} else {
    // Something goes wrong...
    Error Err = JsonOrErr.takeError();
    // Start to handle `Err`...
}
```

The preceding JsonOrErr class has a type of Expected<json::Value>. This means that this Expected variable either carries a json::Value-type Success result or an error, represented by the Error class we just learned about in the previous section.

Just as with the Error class, every Expected instance needs to be *checked*. If it represents an error, that Error instance needs to be *handled* as well. To check the status of an Expected instance, we can also cast it to a Boolean type. However, unlike with Error, if an Expected instance contains a Success result, it will be true after being casted into a Boolean.

If the Expected instance represents a Success result, you can fetch the result using either the * operator (as shown in the preceding code snippet), the -> operator, or the get method. Otherwise, you can retrieve the error by calling the takeError method before handling the Error instance, using the skills we learned in the previous section.

Optionally, if you are sure that an Expected instance is in an Error state, you can check the underlying error type by calling the errorIsA method without retrieving the underlying Error instance first. For example, the following code checks if an error is a FileNotFoundError instance, which we created in the previous section:

```
if (JsonOrErr) {
    // Success!
    ...
} else {
    // Something goes wrong...
    if (JsonOrErr.errorIsA<FileNotFoundError>()) {
        ...
    }
}
```

These are tips for consuming an `Expected` variable. To create an `Expected` instance, the most common way is to leverage the *implicit* type conversion to `Expected`. Here is an example of this:

```
Expected<std::string> readFile(StringRef FileName) {
  if (openFile(FileName)) {
    std::string Content;
    // Reading the file…
    return Content;
  } else
    return make_error<FileNotFoundError>(FileName);
}
```

The preceding code shows that in cases where something goes wrong, we can simply return an `Error` instance, which will be implicitly converted into an `Expected` instance representing that error. Similarly, if everything goes pretty smoothly, the `Success` result—in this case, the `std::string` type variable, `Content`—will also be implicitly converted into an `Expected` instance with a `Success` state.

You have now learned how to use the `Expected` class. The last part of this section will show you how to use one of its sibling classes: `ErrorOr`.

The ErrorOr class

The `ErrorOr` class uses a model that is nearly identical to the `Expected` class— it is either a `Success` result or an error. Unlike the `Expected` class, `ErrorOr` uses `std::error_code` to represent the error. Here is an example of using the `MemoryBuffer` API to read a file—`foo.txt`— and storing its content into a `MemoryBuffer` object:

```
#include "llvm/Support/MemoryBuffer.h"
…
ErrorOr<std::unique_ptr<MemoryBuffer>> ErrOrBuffer
  = MemoryBuffer::getFile("foo.txt");
if (ErrOrBuffer) {
  // Success!
  std::unique_ptr<MemoryBuffer> &MB = *ErrOrBuffer;
} else {
  // Something goes wrong…
  std::error_code EC = ErrOrBuffer.getError();
```

```
    ...
}
```

The previous code snippet shows a similar structure, with the sample code for `Expected` we saw previously: the `std::unique_ptr<MemoryBuffer>` instance is the type of success result here. We can also retrieve it using the `*` operator after checking the state of `ErrOrBuffer`.

The only difference here is that if `ErrOrBuffer` is in an `Error` state, the error is represented by a `std::error_code` instance rather than `Error`. Developers are not *obliged* to handle a `std::error_code` instance—in other words, they can just ignore that error, which might increase the chances of other developers making mistakes in the code. Nevertheless, using the `ErrorOr` class can give you better *interoperability* with C++ standard library APIs, as many of them use `std::error_code` to represent errors. For details about how to use `std::error_code`, please refer to the C++ reference documentation.

Finally, to create an `ErrorOr` instance, we are using the same trick we used on the `Expected` class—leveraging implicit conversion, as shown in the following code snippet:

```
#include <system_error>
ErrorOr<std::string> readFile(StringRef FileName) {
    if (openFile(FileName)) {
        std::string Content;
        // Reading the file...
        return Content;
    } else
        return std::errc::no_such_file_or_directory;
}
```

The `std::errc::no_such_file_or_directory` object is one of the predefined `std::error_code` objects from the `system_error` header file.

In this section, we learned how to use some error-handling utilities provided by LLVM— the important `Error` class that imposes strict rules on unhandled errors, and the `Expected` and `ErrorOr` classes that provide you with a handy way of multiplexing the program result and error state in a single object. These tools can help you to write expressive yet robust error-handling code when developing with LLVM.

Summary

In this chapter, we learned lots of useful utilities that can improve our productivity when developing with LLVM. Some of them—such as optimization remarks or timers—are useful for diagnosing problems raised by LLVM, while others—the `Error` class, for instance—help you to build more robust code that scales well with the complexity of your own compiler.

In the final chapter of this book, we are going to learn about **profile-guided optimization** (**PGO**) and sanitizer development, which are advanced topics that you can't miss.

12
Learning LLVM IR Instrumentation

In the previous chapter, we learned how to leverage various utilities to improve our productivity while developing with LLVM. Those skills can give us a smoother experience when diagnosing problems that are raised by LLVM. Some of these utilities can even reduce the number of potential mistakes that are made by compiler engineers. In this chapter, we are going to learn how instrumentation works in LLVM IR.

The **instrumentation** we are referring to here is a kind of technique that inserts some *probes* into the code we are compiling in order to collect runtime information. For example, we can collect information about how many times a certain function was called – which is only available once the target program has been executed. The advantage of this technique is that it provides extremely accurate information about the target program's behavior. This information can be used in several different ways. For instance, we can use the collected values to compile and optimize the same code *again* – but this time, since we have accurate data, we can perform more aggressive optimizations that couldn't be done previously. This technique is also called **Profile-Guided Optimization** (**PGO**). In another example, will be using the inserted probes to catch undesirable incidents that happened at runtime – buffer overflows, race conditions, and double-free memory, to name a few. The probe that's used for this purpose is also called a **sanitizer**.

To implement instrumentation in LLVM, we not only need the help of LLVM pass, but also the synergy between *multiple* subprojects in LLVM – **Clang**, **LLVM IR Transformation**, and **Compiler-RT**. We already know about the first two from earlier chapters. In this chapter, we are going to introduce Compiler-RT and, more importantly, how can we *combine* these subsystems for the purpose of instrumentation.

Here is the list of topics we are going to cover:

- Developing a sanitizer
- Working with PGO

In the first part of this chapter, we are going to see how a sanitizer is implemented in Clang and LLVM, before creating a simple one by ourselves. The second half of this chapter is going to show you how to use the PGO framework in LLVM and how we can *extend* it.

Technical requirements

In this chapter, we are going to work with multiple subprojects. One of them – Compiler-RT – needs to be included in your build by us modifying the CMake configuration. Please open the CMakeCache.txt file in your build folder and add the compiler-rt string to the value of the LLVM_ENABLE_PROJECTS variable. Here is an example:

```
//Semicolon-separated list of projects to build...
LLVM_ENABLE_PROJECTS:STRING="clang;compiler-rt"
```

After editing the file, launch a build with any build target. CMake will try to reconfigure itself.

Once everything has been set up, we can build the components we need for this chapter. Here is an example command:

```
$ ninja clang compiler-rt opt llvm-profdata
```

This will build the clang tool we're all familiar with and a collection of Compiler-RT libraries, which we are going to introduce shortly.

You can find the sample code for this chapter in the same GitHub repository: https://github.com/PacktPublishing/LLVM-Techniques-Tips-and-Best-Practices-Clang-and-Middle-End-Libraries/tree/main/Chapter12.

Developing a sanitizer

A sanitizer is a kind of technique that checks certain runtime properties of the code (`probe`) that's inserted by the compiler. People usually use a sanitizer to ensure program correctness or enforce security policies. To give you an idea of how a sanitizer works, let's use one of the most popular sanitizers in Clang as an example – the **address sanitizer**.

An example of using an address sanitizer

Let's assume we have some simple C code, such as the following:

```
int main(int argc, char **argv) {
  int buffer[3];
  for (int i = 1; i < argc; ++i)
    buffer[i-1] = atoi(argv[i]);

  for (int i = 1; i < argc; ++i)
    printf("%d ", buffer[i-1]);
  printf("\n");
  return 0;
}
```

The preceding code converted the command-line arguments into integers and stored them in a buffer of size 3. Then, we printed them out.

You should be able to easily spot an outstanding problem: the value of `argc` can be arbitrarily big when it's larger than 3 – the size of `buffer`. Here, we are storing the value in an *invalid* memory location. However, when we compile this code, the compiler will say nothing. Here is an example:

```
$ clang -Wall buffer_overflow.c -o buffer_overflow
$ # No error or warning
```

In the preceding command, even if we enable all the compiler warnings via the `-Wall` flag, `clang` won't complain about the potential bug.

If we try to execute the `buffer_overflow` program, the program will crash at some time point after we pass more than three command-line arguments to it; for example:

```
$ ./buffer_overflow 1 2 3
1 2 3
$ ./buffer_overflow 1 2 3 4
```

```
Segmentation fault (core dumped)
$
```

What's worse, the number of command-line arguments to crash `buffer_overflow` actually *varies* from machine to machine. This makes it even more difficult to debug if the example shown here were a real-world bug. To summarize, the problem we're encountering here is caused by the fact that `buffer_overflow` only goes rogue on *some* inputs and the compiler failed to catch the problem.

Now, let's try to use an address sanitizer to catch this bug. The following command asks `clang` to compile the same code with an address sanitizer:

```
$ clang -fsanitize=address buffer_overflow.c -o san_buffer_
overflow
```

Let's execute the program again. Here is the output:

```
$ ./san_buffer_overflow 1 2 3
1 2 3
$ ./san_buffer_overflow 1 2 3 4
=================================================================
===
==137791==ERROR: AddressSanitizer: stack-buffer-overflow on
address 0x7ffea06bccac at pc 0x0000004f96df bp 0x7ffea06bcc70…
WRITE of size 4 at 0x7ffea06bccac thread T0
…

  This frame has 1 object(s):
    [32, 44) 'buffer' <== Memory access at offset 44 overflows
this variable
…

==137791==ABORTING
$
```

Instead of just crashing, the address sanitizer gave us many details about the issue that was raised at runtime: the sanitizer told us that it detected a *buffer overflow* on the stack, which might be the `buffer` variable.

These messages were extremely useful. Imagine that you are working on a much more complicated software project. When a strange memory bug occurs, rather than just crash or silently change the program's logic, the address sanitizer can point out the problematic area – with high accuracy – right away.

To go a little deeper into its mechanisms, the following diagram illustrates how the address sanitizer detects the buffer overflow:

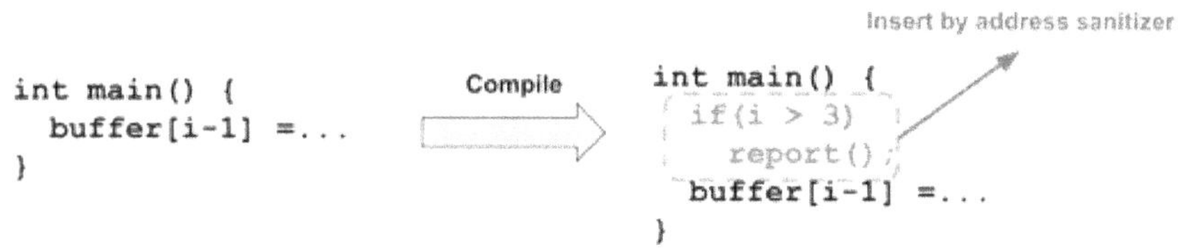

Figure 12.1 – Instrumentation code inserted by the address sanitizer

Here, we can see that the address sanitizer is effectively inserting a boundary check into the array index that's used for accessing `buffer`. With this extra check – which will be executed at runtime – the target program can bail out with error details before violating the memory access. More generally speaking, during the compilation, a sanitizer inserts some instrumentation code (into the target program) that will eventually be executed at runtime to check or *guard* certain properties.

Detecting overflow using an address sanitizer

The preceding diagram shows a simplified version of how an address sanitizer works. In reality, the address sanitizer will leverage multiple strategies to monitor memory access in a program. For example, an address sanitizer can use a special memory allocator that allocates memory with `traps` put at the invalid memory region.

While an address sanitizer is specialized in catching illegal memory access, a **ThreadSanitizer** can be used to catch data race conditions; that is, invalid access from multiple threads on the same chunk of data. Some other examples of sanitizers in Clang are the **LeakSanitizer**, which is used for detecting sensitive data such as passwords being leaked, and **MemorySanitizer**, which is used for detecting reads to uninitialized memory.

Of course, there are some downsides to using sanitizers. The most prominent problem is the performance impact: using a thread sanitizer (in Clang) as an example, programs that are compiled with one are *5~15 times slower* than the original version. Also, since sanitizers insert extra code into the program, it might hinder some optimization opportunities, or even affect the original program's logic! In other words, it is a trade-off between the *robustness* and *performance* of the target program.

With that, you've learned about the high-level idea of a sanitizer. Let's try to create a real one by ourselves to understand how Clang and LLVM implement a sanitizer. The following section contains more code than any of the examples in previous chapters, not to mention the changes are spread across different subprojects in LLVM. To focus on the most important knowledge, we won't go into the details of some *supporting* code – for example, changes that are made to CMake build scripts. Instead, we will go through them by providing a brief introduction and pointing out where you can find it in this book's GitHub repository.

Let's start by providing an overview of the project we are going to create.

Creating a loop counter sanitizer

To (slightly) simplify our task, the sanitizer we are going to create – a loop counter sanitizer, or **LPCSan** for short – looks just like a sanitizer except that it is not checking any serious program properties. Instead, we want to use it to print out the real, concrete **trip count** – the number of iterations – of a loop, which is only available during runtime.

For example, let's assume we have the following input code:

```
void foo(int S, int E, int ST, int *a) {
  for (int i = S; i < E; i += ST) {
    a[i] = a[i + 1];
  }
}
int main(int argc, char **argv) {
  int start = atoi(argv[1]),
      end = atoi(argv[2]),
      step = atoi(argv[3]);
  int a[100];
  foo(start, end, step, a);
  return 0;
}
```

We can compile it with a LPCSan using the following command:

```
$ clang -O1 -fsanitize=loop-counter test_lpcsan.c -o test_
lpcsan
```

Note that compiling with optimization greater than -O0 is necessary; we will explain why later.

When we execute `test_lpcsan` (with some command-line argument), we can print out the precise trip count of the loop in the `foo` function. For example, look at the following code:

```
$ ./test_lpcsan 0 100 1
==143813==INFO: Found a loop with trip count 100
$ ./test_lpcsan 0 50 2
==143814==INFO: Found a loop with trip count 25
$
```

The message highlighted in the preceding code was printed by our sanitizer code.

Now, let's dive into the steps for creating the LPCSan. We will divide this tutorial into three parts:

- Developing an IR transformation
- Adding Compiler-RT components
- Adding the LPCSan to Clang

We will start with the IR transformation part of this sanitizer.

Developing an IR transformation

Previously, we learned that an address sanitizer – or just a sanitizer in general – usually inserts code into the target program to check certain runtime properties or collect data. In *Chapter 9, Working with PassManager and AnalysisManager*, and *Chapter 10, Processing LLVM IR*, we learned how to modify/transform LLVM IR, including inserting new code into it, so this seems to be a good starting point for crafting our LPCSan.

In this section, we are going to develop an LLVM pass called `LoopCounterSanitizer` that inserts special function calls to collect the exact trip count of every loop in `Module`. Here are the detailed steps:

1. First, let's create two files: `LoopCounterSanitizer.cpp` under the `llvm/lib/Transforms/Instrumentation` folder and its corresponding header file inside the `llvm/include/llvm/Transforms/Instrumentation` folder. Inside the header file, we will place the declaration of this pass, as shown here:

    ```
    struct LoopCounterSanitizer
        : public PassInfoMixin<LoopCounterSanitizer> {
      PreservedAnalyses run(Loop&, LoopAnalysisManager&,
                            LoopStandardAnalysisResults&,
                            LPMUpdater&);
    ```

```
private:
  // Sanitizer functions
  FunctionCallee LPCSetStartFn, LPCAtEndFn;
  void initializeSanitizerFuncs(Loop&);
};
```

The preceding code shows the typical loop pass structure we saw in *Chapter 10, Processing LLVM IR*. The only notable changes are the LPCSetStartFn and LPCAtEndFn memory variables – they will store the Function instances that collect loop trip counts (FunctionCallee is a thin wrapper around Function that provides additional function signature information).

2. Finally, in LoopCounterSanitizer.cpp, we are placing the skeleton code for our pass, as shown here:

```
PreservedAnalyses
LoopCounterSanitizer::run(Loop &LP, LoopAnalysisManager
&LAM, LoopStandardAnalysisResults &LSR, LPMUpdater &U) {
  initializeSanitizerFuncs(LP);
  return PreservedAnalyses::all();
}
```

The initializeSanitizerFuncs method in the preceding code will populate LPCSetStartFn and LPCAtEndFn. Before we go into the details of initializeSanitizerFuncs, let's talk more about LPCSetStartFn and LPCAtEndFn.

3. To figure out the exact trip count, the Function instance stored in LPCSetStartFn will be used to collect the *initial* induction variable value of a loop. On the other hand, the Function instance stored in LPCAtEndFn will be used to collect the *final* induction variable value and the step value of the loop. To give you a concrete idea of how these two Function instances work together, let's assume we have the following pseudocode as our input program:

```
void foo(int S, int E, int ST) {
  for (int i = S; i < E; i += ST) {
    ...
  }
}
```

In the preceding code, the S, E, and ST variables represent the initial, final, and step values of a loop, respectively. The goal of the LoopCounterSanitizer pass is to insert LPCSetStartFn and LPCAtEndFn in the following way:

```
void foo(int S, int E, int ST) {
    for (int i = S; i < E; i += ST) {
        lpc_set_start(S);
        …
        lpc_at_end(E, ST);
    }
}
```

lpc_set_start and lpc_at_end in the preceding code are Function instances that are stored in LPCSetStartFn and LPCAtEndFn, respectively. Here is one of the possible (pseudo) implementations of these two functions:

```
static int CurrentStartVal = 0;
void lpc_set_start(int start) {
    CurrentStartVal = start;
}
void lpc_at_end(int end, int step) {
    int trip_count = (end - CurrentStartVal) / step;
    printf("Found a loop with trip count %d\n",
      trip_count);
}
```

Now that we know the roles of LPCSetStartFn and LPCAtEndFn, it's time to take a look at how initializeSanitizerFuncs initializes them.

4. Here is the code inside initializeSanitizerFuncs:

```
void LoopCounterSanitizer::initializeSanitizerFuncs(Loop
&LP) {
    Module &M = *LP.getHeader()->getModule();
    auto &Ctx = M.getContext();
    Type *VoidTy = Type::getVoidTy(Ctx),
        *ArgTy = Type::getInt32Ty(Ctx);
    LPCSetStartFn
      = M.getOrInsertFunction("__lpcsan_set_loop_start",
                              VoidTy, ArgTy);
```

```
LPCAtEndFn = M.getOrInsertFunction("__lpcsan_at_loop_
  end", VoidTy, ArgTy, ArgTy);
}
```

The previous code is basically fetching two functions, `__lpcsan_set_loop_start` and `__lpcsan_at_loop_end`, from the module and storing their `Function` instances in `LPCSetStartFn` and `LPCAtEndFn`, respectively.

The `Module::getOrInsertFunction` method either grabs the `Function` instance of the given function name from the module or creates one if it doesn't exist. If it's a newly created instance, it has an empty function body; in other words, it only has a function *declaration*.

It is also worth noting that the second argument of `Module::getOrInsertFunction` is the return type of the `Function` inquiry. The rest (the arguments for `getOrInsertFunction`) represent the argument types of that `Function`.

With `LPCSetStartFn` and `LPCAtEndFn` set up, let's see how we can insert them into the right place in IR.

5. Recall that in *Chapter 10*, *Processing LLVM IR*, we learned about several utility classes for working with `Loop`. One of them – `LoopBounds` – can give us the boundary of a `Loop`. We can do this by including the start, end, and step values of an induction variable, which is exactly the information we are looking for. Here is the code that tries to retrieve a `LoopBounds` instance:

```
PreservedAnalyses
LoopCounterSanitizer::run(Loop &LP, LoopAnalysisManager
&LAM, LoopStandardAnalysisResults &LSR, LPMUpdater &U) {
  initializeSanitizerFuncs(LP);
  ScalarEvolution &SE = LSR.SE;

  using LoopBounds = typename Loop::LoopBounds;
  auto MaybeLB = LP.getBounds(SE);
  if (!MaybeLB) {
    errs() << "WARNING: Failed to get loop bounds\n";
    return PreservedAnalyses::all();
  }
  LoopBounds &LB = *MaybeLB;
  ...
```

```
    Value *StartVal = &LB.getInitialIVValue(),
           *EndVal = &LB.getFinalIVValue(),
           *StepVal = LB.getStepValue();
}
```

`Loop::getBounds` from the preceding code returned an `Optional<LoopBounds>` instance. The `Optional<T>` class is a useful container that either stores an instance of the `T` type or is *empty*. You can think of it as a replacement for the **null pointer**: usually, people use `T*` to represent a computation result where a null pointer means an empty value. However, this has the risk of dereferencing a null pointer if the programmer forgets to check the pointer first. The `Optional<T>` class doesn't have this problem.

With a `LoopBounds` instance, we can retrieve the induction variable's range and store it in the `StartVal`, `EndVal`, and `StepVal` variables.

6. `StartVal` is the `Value` instance to be collected by `__lpcsan_set_loop_start`, whereas `__lpcsan_at_loop_end` is going to collect `EndVal` and `StepVal` at runtime. Now, the question is, *where* should we insert function calls to `__lpcsan_set_loop_start` and `__lpcsan_at_loop_end` to correctly collect those values?

The rule of thumb is that we need to insert those function calls after the *definition* of those values. While we can find the exact locations where those values were defined, let's try to simplify the problem by inserting instrumentation function calls at some fixed locations – locations where our target values are *always* available.

For `__lpcsan_set_loop_start`, we are inserting it at the end of the **loop header** block, because the initial induction variable value will never be defined after this block. Here is the code:

```
// Inside LoopCounterSanitizer::run …
…
BasicBlock *Header = LP.getHeader();
Instruction *LastInst = Header->getTerminator();
IRBuilder<> Builder(LastInst);
Type *ArgTy = LPCSetStartFn.getFunctionType()-
>getParamType(0);

if (StartVal->getType() != ArgTy) {
  // cast to argument type first
  StartVal = Builder.CreateIntCast(StartVal, ArgTy,
```

```
  true);
}
Builder.CreateCall(LPCSetStartFn, {StartVal});
...
```

In the preceding code, we used `getTerminator` to get the last `Instruction` from the header block. Then, we used `IRBuilder<>` – with the last instruction as the insertion point – to insert new `Instruction` instances.

Before we can pass `StartVal` as an argument to the new `__lpcsan_set_loop_start` function call, we need to convert its IR type (represented by the `Type` class) into a compatible one. `IRBuilder::CreateInstCast` is a handy utility that automatically generates either an instruction to *extend* the integer bit width or an instruction to *truncate* the bit width, depending on the given `Value` and `Type` instances.

Finally, we can create a function call to `__lpcsan_set_loop_start` via `IRBuilder::CreateCall`, with `StartVal` as the function call argument.

7. For `__lpcsan_at_loop_end`, we are using the same trick to collect the runtime values of `EndVal` and `StepVal`. Here is the code:

```
BasicBlock *ExitBlock = LP.getExitBlock();
Instruction *FirstInst = ExitBlock->getFirstNonPHI();
IRBuilder<> Builder(FirstInst);
FunctionType *LPCAtEndTy = LPCAtEndFn.getFunctionType();
Type *EndArgTy = LPCAtEndTy->getParamType(0),
     *StepArgTy = LPCAtEndTy->getParamType(1);

if (EndVal->getType() != EndArgTy)
  EndVal = Builder.CreateIntCast(EndVal, EndArgTy, true);
if (StepVal->getType() != StepArgTy)
  StepVal = Builder.CreateIntCast(StepVal, StepArgTy,
    true);

Builder.CreateCall(LPCAtEndFn, {EndVal, StepVal});
```

Different from the previous step, we are inserting the function call to `__lpcsan_` `at_loop_end` at the beginning of the *exit block*. This is because we can always expect the end value and the step value of the induction variable being defined before we leave the loop.

These are all the implementation details for the `LoopCounterSanitizer` pass.

8. Before we wrap up this section, we need to edit a few more files to make sure everything works. Please look at the `Changes-LLVM.diff` file in the sample code folder for this chapter. Here is the summary of the changes that were made in other supporting files:

    i. Changes in `llvm/lib/Transforms/Instrumentation/CMakeLists.txt`: Add our new pass source file to the build.

    ii. Changes in `llvm/lib/Passes/PassRegistry.def`: Add our pass to the list of available passes so that we can test it using our old friend `opt`.

With that, we've finally finished making all the necessary modifications to the LLVM part.

Before we move on to the next section, let's test our newly created `LoopCounterSanitizer` pass. We are going to be using the same C code we saw earlier in this section. Here is the function that contains the loop we want to instrument:

```
void foo(int S, int E, int ST, int *a) {
   for (int i = S; i < E; i += ST) {
     a[i] = a[i + 1];
   }
}
```

Note that although we didn't explicitly check the loop form in our pass, some of the APIs that were used in the pass actually required the loop to be *rotated*, so please generate the LLVM IR code with an O1 optimization level to make sure the loop rotation's Pass has kicked in:

Here is the simplified LLVM IR for the `foo` function:

```
define void @foo(i32 %S, i32 %E, i32 %ST, i32* %a) {
  %cmp9 = icmp slt i32 %S, %E
  br i1 %cmp9, label %for.body.preheader, label %for.cond.
  cleanup

for.body.preheader:
  %0 = sext i32 %S to i64
```

```
%1 = sext i32 %ST to i64
%2 = sext i32 %E to i64
br label %for.body
...
for.body:
  %indvars.iv = phi i64 [ %0, %for.body.preheader ], [
  %indvars.iv.next, %for.body ]
  ...
  %indvars.iv.next = add i64 %indvars.iv, %1
  %cmp = icmp slt i64 %indvars.iv.next, %2
  br i1 %cmp, label %for.body, label %for.cond.cleanup
}
```

The highlighted labels are the preheader and loop body blocks for this loop. Since this loop has been rotated, the for.body block is both the header, latch, and exiting block for this loop.

Now, let's transform this IR with opt using the following command:

```
$ opt -S -passes="loop(lpcsan)" input.ll -o -
```

In the –passes command-line option, we asked opt to run our LoopCounterSanitizer pass (with the name lpcsan, which is registered in the PassRegistry.def file). The enclosing loop(...) string is simply telling opt that lpcsan is a loop pass (you can actually omit this decoration since opt can find the right pass most of the time).

Here is the simplified result:

```
declare void @__lpcsan_set_loop_start(i32)
declare void @__lpcsan_at_loop_end(i32, i32)

define void @foo(i32 %S, i32 %E, i32* %a) {
  %cmp8 = icmp slt i32 %S, %E
  br i1 %cmp8, label %for.body.preheader, label %for.cond.
cleanup

for.body.preheader:
  %0 = sext i32 %S to i64
  %wide.trip.count = sext i32 %E to i64
```

```
      br label %for.body

for.cond.cleanup.loopexit:
  %1 = trunc i64 %wide.trip.count to i32
  call void @__lpcsan_at_loop_end(i32 %1, i32 1)
  br label %for.cond.cleanup

for.body:
  ...
  %3 = trunc i64 %0 to i32
  call void @__lpcsan_set_loop_start(i32 %3)
  br i1 %exitcond.not, label %for.cond.cleanup.loopexit, label
    %for.body
}
```

As you can see, `__lpcsan_set_loop_start` and `__lpcsan_at_loop_end` have been correctly inserted into the header block and exit block, respectively. They are also collecting the desired values related to the loop trip count.

Now, the biggest question is: where are the *function bodies* for `__lpcsan_set_loop_start` and `__lpcsan_at_loop_end`? Both only have declarations in the preceding IR code.

In the next section, we will use Compiler-RT to answer this question.

Adding the Compiler-RT component

The name **Compiler-RT** stands for **Compiler RunTime**. The usage of *runtime* is a little ambiguous here because too many things can be called a runtime in a normal compilation pipeline. But the truth is that Compiler-RT *does* contain a wide range of libraries for completely different tasks. What these libraries have in common is that they provide *supplement* code for the target program to implement enhancement features or functionalities that were otherwise absent. It is important to remember that Compiler-RT libraries are NOT used for building a compiler or related tool – they should be linked with the program we are compiling.

One of the most used features in Compiler-RT is the **builtin function**. As you might have heard, more and more computer architectures nowadays support *vector operation* natively. That is, you can process multiple data elements at the same time with the support from hardware. Here is some example code, written in C, that uses vector operations:

```
typedef int v4si __attribute__((__vector_size__(16)));
v4si v1 = (v4si){1, 2, 3, 4};
v4si v2 = (v4si){5, 6, 7, 8};
v4si v3 = v1 + v2; // = {6, 8, 10, 12}
```

The preceding code used a non-standardized (currently, you can only use this syntax in Clang and GCC) C/C++ vector extension to declare two vectors, v1 and v2, before adding them to yield a third one.

On X86-64 platforms, this code will be compiled to use one of the vector instruction sets, such as **SSE** or **AVX**. On the ARM platform, the resulting binary might be using the **NEON** vector instruction set. But what if your target platform does NOT have a vector instruction set? The most obvious solution would be "synthesizing" these unsupported operations with the available instructions. For example, we should write a for-loop to replace vector summation in this case. More specifically, whenever we see a vector summation at compilation time, we replace it with a call to a function that contains the synthesis implementation using for-loop. The function body can be put anywhere, as long as it is eventually linked with the program. The following diagram illustrates this process:

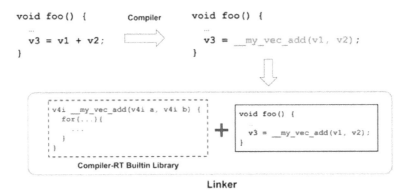

Figure 12.2 – Workflow of the Compiler-RT builtin

As you may have noticed, the workflow shown here is similar to our requirement in the LPCSan: in the previous section, we developed an LLVM pass that inserted extra function calls to collect the loop trip count, but we still need to implement those collector functions. If we leverage the workflow shown in the preceding diagram, we can come up with a design, as shown in the following diagram:

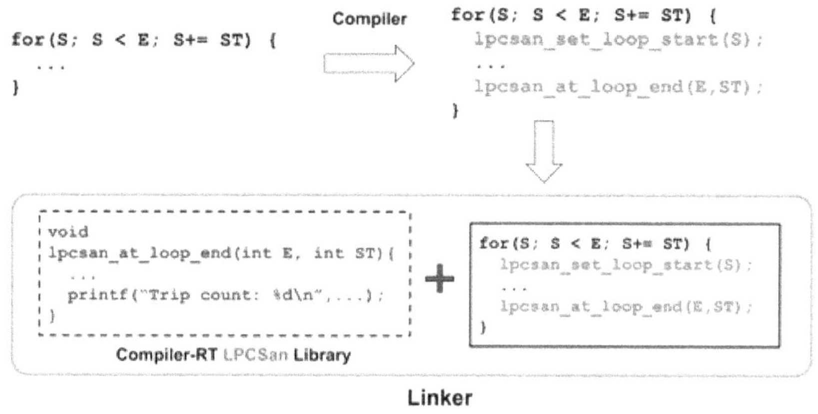

Figure 12.3 – Workflow of the Compiler-RT LPCSan component

The previous diagram shows that the function bodies of `__lpcsan_set_loop_start` and `__lpcsan_at_loop_end` are put inside a Compiler-RT library that will eventually be linked with the final binary. Inside these two functions, we calculate the trip count using the input arguments and print the result. In the rest of this section, we'll show you how to create such a Compiler-RT library for the LPCSan. Let's get started:

1. First, switch the folder to `llvm-project/compiler-rt`, the root of Compiler-RT. Inside this subproject, we must create a new folder called `lib/lpcsan` before we put a new `lpcsan.cpp` file inside it. Within this file, let's create the skeleton for our instrumentation functions. Here is the code:

    ```cpp
    #include "sanitizer_common/sanitizer_common.h"
    #include "sanitizer_common/sanitizer_internal_defs.h"
    using namespace __sanitizer;

    extern "C" SANITIZER_INTERFACE_ATTRIBUTE
    void __lpcsan_set_loop_start(s32 start){
      // TODO
    }

    extern "C" SANITIZER_INTERFACE_ATTRIBUTE
    void __lpcsan_at_loop_end(s32 end, s32 step){
      // TODO
    }
    ```

There are two things worth noting here: first, use the primitive data types provided by Compiler-RT. For example, in the preceding code, we used s32 – available under the __sanitizer namespace – for a signed 32-bit integer rather than the normal int. The rationale behind this is that we might need to build Compiler-RT libraries for different hardware architectures or platforms, and the width of int might not be 32 bits on some of them.

Second, although we are using C++ to implement our instrumentation functions, we need to expose them as C functions because C functions have a more stable **Application Binary Interface** (**ABI**). Therefore, please make sure to add extern "C" to functions you want to export. The SANITIZER_INTERFACE_ATTRIBUTE macro also ensures that the function will be exposed at the library interface correctly, so please add this as well.

2. Next, we will add the necessary code to these two functions. Here is how we do this:

```
static s32 CurLoopStart = 0;

extern "C" SANITIZER_INTERFACE_ATTRIBUTE
void __lpcsan_set_loop_start(s32 start){
  CurLoopStart = start;
}

extern "C" SANITIZER_INTERFACE_ATTRIBUTE
void __lpcsan_at_loop_end(s32 end, s32 step){
  s32 trip_count = (end - CurLoopStart) / step;
  s32 abs_trip_count
    = trip_count >= 0? trip_count : -trip_count;
  Report("INFO: Found a loop with "
         "trip count %d\n", abs_trip_count);
}
```

The implementation we used here is pretty straightforward: CurLoopStart is a global variable that memorizes the *initial* induction variable value of the current loop. This is updated by __lpcsan_set_loop_start.

Recall that when a loop is complete, __lpcsan_at_loop_end will be invoked. When that happens, we use the value stored in CurLoopStart and the end and step arguments to calculate the exact trip count of the current loop, before printing the result.

3. Now that we have implemented the core logic, it's time to build this library. Inside the `lib/lpcsan` folder, create a new `CMakeLists.txt` file and insert the following code:

```
...
set(LPCSAN_RTL_SOURCES
    lpcsan.cpp)
add_compiler_rt_component(lpcsan)

foreach(arch ${LPCSAN_SUPPORTED_ARCH})
  set(LPCSAN_CFLAGS ${LPCSAN_COMMON_CFLAGS})
  add_compiler_rt_runtime(clang_rt.lpcsan
    STATIC
    ARCHS ${arch}
    SOURCES ${LPCSAN_RTL_SOURCES}
        $<TARGET_OBJECTS:RTSanitizerCommon.${arch}>
        $<TARGET_OBJECTS:RTSanitizerCommonLibc.${arch}>
    ADDITIONAL_HEADERS ${LPCSAN_RTL_HEADERS}
    CFLAGS ${LPCSAN_CFLAGS}
    PARENT_TARGET lpcsan)
  ...
endforeach()
```

In the preceding code, we are only showing the most important part of our `CMakeLists.txt`. Here are some highlights:

i. Compiler-RT creates its own set of CMake macros/functions. Here, we are using two of them, `add_compiler_rt_component` and `add_compiler_rt_runtime`, to create a pseudo build target for the entire LPCSan and the real library build target, respectively.

ii. Different from a conventional build target, if a sanitizer wants to use supporting/utility libraries in Compiler-RT – for example, `RTSanitizerCommon` in the preceding code – we usually link against their *object files* rather than their library files. More specifically, we can use the `$<TARGET_OBJECTS:...>` directive to import supporting/utility components as one of the input sources.

iii. A sanitizer library can support multiple architectures and platforms. In Compiler-RT, we are enumerating all the supported architectures and creating a sanitizer library for each of them.

Again, the preceding snippet is just a small part of our build script. Please refer to our sample code folder for the complete `CMakeLists.txt` file.

4. To successfully build the LPCSan, we still need to make some changes in Compiler-RT. The `Base-CompilerRT.diff` patch in the same code folder provides the rest of the changes that are necessary to build our sanitizer. Apply it to Compiler-RT's source tree. Here is the summary of this patch:

   i. Changes in `compiler-rt/cmake/config-ix.cmake` basically specify the supported architectures and operating systems of the LPCSan. The `LPCSAN_SUPPORTED_ARCH` CMake variable we saw in the previous snippet comes from here.

   ii. The entire `compiler-rt/test/lpcsan` folder is actually a placeholder. For some reason, having tests is a *requirement* for every sanitizer in Compiler-RT – which is different from LLVM. Therefore, we are putting an empty test folder here to pass this requirement that's being imposed by the build infrastructure.

These are all the steps for producing a Compiler-RT component for our LPCSan.

To just build our LPCSan library, invoke the following command:

```
$ ninja lpcsan
```

Unfortunately, we can't test this LPCSan library until we've modified the compilation pipeline in Clang. In the last part of this section, we are going to learn how to achieve this task.

Adding the LPCSan to Clang

In the previous section, we learned how Compiler-RT libraries provide supplement functionalities to the target program or assist with special instrumentation, such as the sanitizer we just created. In this section, we are going to put everything together so that we can use our LPCSan simply by passing the `-fsanitize=loop-counter` flag to `clang`.

Recall that in *Figure 12.3*, Compiler-RT libraries need to be linked with the program we are compiling. Also, recall that in order to insert the instrumentation code into the target program, we must run our `LoopCounterSanitizer` pass. In this section, we are going to modify the compilation pipeline in Clang so that it runs our LLVM pass at a certain time and sets up the correct configuration for our Compiler-RT library. More specifically, the following diagram shows the tasks that each component needs to complete to run our LPCSan:

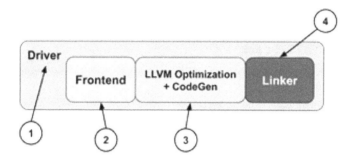

Figure 12.4 – Tasks for each component in the pipeline

Here are the descriptions for each of the numbers (enclosed in circles) in the preceding diagram:

1. The driver needs to recognize the `-fsanitize=loop-counter` flag.

2. When the frontend is about to generate LLVM IR from an **Abstract Syntax Tree (AST)**, it needs to correctly configure the LLVM pass pipeline so that it includes the `LoopCounterSanitizer` pass.

3. The LLVM pass pipeline needs to run our `LoopCounterSanitizer` (we don't need to worry about this task if the previous task is done correctly).

4. The linker needs to link our Compiler-RT library to the target program.

Although this workflow looks a little scary, don't be overwhelmed by the prospective workload – Clang can actually do most of these tasks for you, as long as you provide sufficient information. In the rest of this section, we'll show you how to implement the tasks shown in the preceding diagram to fully integrate our LPCSan into the Clang compilation pipeline (the following tutorial works inside the `llvm-project/clang` folder). Let's get started:

1. First, we must modify `include/clang/Basic/Sanitizers.def` to add our sanitizer:

    ```
    ...
    // Shadow Call Stack
    SANITIZER("shadow-call-stack", ShadowCallStack)

    // Loop Counter Sanitizer
    SANITIZER("loop-counter", LoopCounter)
    ...
    ```

    This effectively adds a new enum value, `LoopCounter`, to the `SanitizerKind` class.

    It turns out that the driver will parse the `-fsanitize` command-line option and *automatically* translate `loop-counter` into `SanitizerKind::LoopCounter` based on the information we provided in `Sanitizers.def`.

2. Next, let's work on the driver part. Open `include/clang/Driver/SanitizerArgs.h` and add a new utility method, `needsLpcsanRt`, to the `SanitizerArgs` class. Here is the code:

    ```
    bool needsLsanRt() const {…}
    bool needsLpcsanRt() const {
       return Sanitizers.has(SanitizerKind::LoopCounter);
    }
    ```

    The utility method we created here can be used by other places in the driver to check if our sanitizer needs a Compiler-RT component.

3. Now, let's navigate to the `lib/Driver/ToolChains/CommonArgs.cpp` file. Here, we're adding a few lines to the `collectSanitizerRuntimes` function. Here is the code:

```
...
if (SanArgs.needsLsanRt() && SanArgs.linkRuntimes())
  StaticRuntimes.push_back("lsan");
if (SanArgs.needsLpcsanRt() && SanArgs.linkRuntimes())
  StaticRuntimes.push_back("lpcsan");
...
```

The preceding snippet effectively makes the linker link the correct Compiler-RT library to the target binary.

4. The last change we will make to the driver is in `lib/Driver/ToolChains/Linux.cpp`. Here, we add the following lines to the `Linux::getSupportedSanitizers` method:

```
SanitizerMask Res = ToolChain::getSupportedSanitizers();
...
Res |= SanitizerKind::LoopCounter;
...
```

The previous code is essentially telling the driver that we support the LPCSan in the current toolchain – the toolchain for Linux. Note that to simplify our example, we are only supporting the LPCSan in Linux. If you want to support this custom sanitizer in other platforms and architectures, modify the other toolchain implementations. Please refer to *Chapter 8, Working with Compiler Flags and Toolchains*, for more details if needed.

5. Finally, we are going to insert our `LoopCounterSanitizer` pass into the LLVM pass pipeline. Open `lib/CodeGen/BackendUtil.cpp` and add the following lines to the `addSanitizers` function:

```
...
// `PB` has the type of `PassBuilder`
PB.registerOptimizerLastEPCallback(
  [&](ModulePassManager &MPM,
      PassBuilder::OptimizationLevel Level) {
    ...
    if (LangOpts.Sanitize.has(SanitizerKind::LoopCounter))
```

```
{
    auto FA
        =
createFunctionToLoopPassAdaptor(LoopCounterSanitizer());
    MPM.addPass(

createModuleToFunctionPassAdaptor(std::move(FA)));
    }
  });
...
```

The enclosing folder for this file, CodeGen, is a place where the Clang and LLVM libraries meet. Therefore, we will see several LLVM APIs appear in this place. There are primarily two tasks for this CodeGen component:

a. Converting the Clang AST into its equivalent LLVM IR module

b. Constructing an LLVM pass pipeline to optimize the IR and generate machine code

The previous snippet was trying to customize the second task – that is, customizing the LLVM Pass pipeline. The specific function – addSanitizers – we are modifying here is responsible for putting sanitizer passes into the pass pipeline. To have a better understanding of this code, let's focus on two of its components:

i. PassBuilder: This class provides predefined pass pipeline configurations for each optimization level – that is, the O0 ~ O3 notations (as well as Os and Oz for size optimization) we are familiar with. In addition to these predefined layouts, developers are free to customize the pipeline by leveraging the **extension point (EP)**.

An EP is a certain position in the (predefined) pass pipeline where you can insert new passes. Currently, PassBuilder supports several EPs, such as at the *beginning* of the pipeline, at the *end* of the pipeline, or at the end of the vectorization process, to name a few. An example of using EP can be found in the preceding code, where we used the PassBuilder::registerOptimizerLastEPCallback method and a lambda function to customize the EP located at the *end* of the Pass pipeline. The lambda function has two arguments: ModulePassManager – which represents the pass pipeline – and the current optimization level. Developers can use ModulePassManager::addPass to insert arbitrary LLVM passes into this EP.

ii. ModulePassManager: This class represents a Pass pipeline – or, more specifically, the pipeline for Module. There are, of course, other PassManager classes for different IR units, such as FunctionPassManager for Function.

In the preceding code, we were trying to use the `ModulePassManager` instance to insert our `LoopCounterSanitizer` pass whenever `SanitizerKind::LoopCounter` was one of the sanitizers that had been designated by the user. Since `LoopCounterSanitizer` is a loop pass rather than a module pass, we need to add some *adaptors* between the pass and PassManager. The `createFunctionToLoopPassAdaptor` and `createModuleToFunctionPassAdaptor` functions we were using here created a special instance that adapts a pass to a PassManager of a different IR unit.

This is all the program logic that supports our LPCSan in the Clang compilation pipeline.

6. Last but not least, we must make a small modification to the build system. Open the `runtime/CMakeLists.txt` file and change the following CMake variable:

```
...
set(COMPILER_RT_RUNTIMES fuzzer asan builtins … lpcsan)
foreach(runtime ${COMPILER_RT_RUNTIMES})
...
```

The change we made to `COMPILER_RT_RUNTIMES` effectively imports our LPCSan Compiler-RT libraries into the build.

These are all the steps necessary to support the LPCSan in Clang. Now, we can finally use the LPCSan in the same way we showed you at the beginning of this section:

```
$ clang -O1 -fsanitize=loop-counter input.c -o input
```

In this section, we learned how to create a sanitizer. A sanitizer is a useful tool for capturing runtime behaviors without modifying the original program code. The ability to create a sanitizer increases the flexibility for compiler developers to create custom diagnosing tools tailored for their own use cases. Developing a sanitizer requires comprehensive knowledge of Clang, LLVM, and Compiler-RT: creating a new LLVM pass, making a new Compiler-RT component, and customizing the compilation pipeline in Clang. You can use the content in this section, to reinforce what you've learned in previous chapters of this book.

In the last section of this chapter, we are going to look at one more instrumentation technique: **PGO**.

Working with PGO

In the previous section, we learned how a sanitizer assists developers in performing sanity checks with higher precision using data that is only available at runtime. We also learned how to create a custom sanitizer. In this section, we will follow up on the idea of leveraging runtime data. We are going to learn an alternative use for such information – using it for compiler optimization.

PGO is a technique that uses statistics that have been collected during runtime to enable more aggressive compiler optimizations. The *profile* in its name refers to the runtime data that's been collected. To give you an idea of how such data enhances an optimization, let's assume we have the following C code:

```
void foo(int N) {
  if (N > 100)
    bar();
  else
    zoo();
}
```

In this code, we have three functions: foo, bar, and zoo. The first function conditionally calls the latter two.

When we try to optimize this code, the optimizer usually tries to *inline* callee functions into the caller. In this case, bar or zoo might be inlined into foo. However, if either bar or zoo has a large function body, inlining *both* might bloat the size of the final binary. Ideally, it will be great if we could inline only the one that executes the most frequently. Sadly, from a statistics point of view, we have no clue about which function has the highest execution frequency, because the foo function conditionally calls either of them based on a (non-constant) variable.

With PGO, we can collect the execution frequencies of both bar and zoo at runtime and use the data to compile (and optimize) the same code *again*. The following diagram shows the high-level overview of this idea:

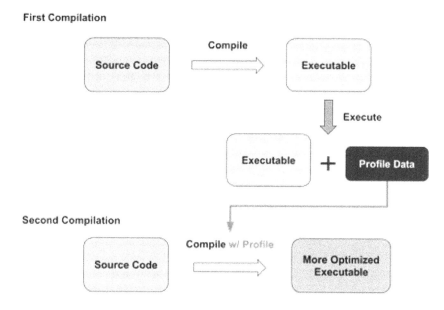

Figure 12.5 – PGO workflow

Here, the first compilation phase compiled and optimized the code normally. After we executed the compiled program (an arbitrary number of times), we were able to collect the profile data files. In the second compilation phase, we not only optimized the code, as we did previously, but also integrated the profile data into the optimizations to make them act more aggressively.

There are primarily two ways for PGO to collect runtime profiling data: inserting **instrumentation** code or leveraging **sampling** data. Let's introduce both.

Introduction to instrumentation-based PGO

Instrumentation-based PGO inserts instrumentation code into the target program during the first compilation phase. This code measures the execution frequency of the program constructions we're interested in – for example, basic blocks and functions – and writes the result in a file. This is similar to how a sanitizer works.

Instrumentation-based PGO usually generates profiling data with higher precision. This is because the compiler can insert instrumentation code in a way that provides the greatest benefit for other optimizations. However, just like the sanitizer, instrumentation-based PGO changes the execution flow of the target program, which increases the risk of performance regression (for the binary that was generated from the first compilation phase).

Introduction to sampling-based PGO

Sampling-based PGO uses *external* tools to collect profiling data. Developers use profilers such as `perf` or `valgrind` to diagnose performance issues. These tools usually leverage advanced system features or even hardware features to collect the runtime behavior of a program. For example, `perf` can give you insights into branch prediction and cache line misses.

Since we are leveraging data from other tools, there is no need to modify the original code to collect profiles. Therefore, sampling-based PGO usually has an extremely low runtime overhead (usually, this is less than 1%). Also, we don't need to recompile the code for profiling purposes. However, profiling data that's generated in this way is usually less precise. It's also more difficult to map the profiling data back to the original code during the second compilation phase.

In the rest of this section, we are going to focus on instrumentation-based PGO. We are going to learn how to leverage it with LLVM IR. Nevertheless, as we will see shortly, these two PGO strategies in LLVM share lots of common infrastructures, so the code is portable. Here is the list of topics we are going to cover:

- Working with profiling data
- Learning about the APIs for accessing profiling data

The first topic will show us how to create and use instrumentation-based PGO profiles with Clang, as well as some of the tools that can help us inspect and modify profiling data. The second topic will give you more details on how to access profiling data using LLVM APIs. This is useful if you want to create your own PGO pass.

Working with profiling data

In this section, we are going to learn how to use generate, inspect, and even modify instrumentation-based profiling data. Let's start with the following example:

```
__attribute__((noinline))
void foo(int x) {
  if (get_random() > 5)
    printf("Hello %d\n", x * 3);
}
int main(int argc, char **argv) {
  for (int i = 0; i < argc + 10; ++i) {
    foo(i);
```

```
    }
    return 0;
}
```

In the preceding code, get_random is a function that generates a random number from 1 to 10 with uniform distribution. In other words, the highlighted if statement in the foo function should have a 50% chance of being taken. In addition to the foo function, the trip count of the for loop within main depends on the number of command-line arguments there are.

Now, let's try to build this code with instrumentation-based PGO. Here are the steps:

1. The first thing we are going to do is generate an executable for PGO profiling. Here is the command:

```
$ clang -O1 -fprofile-generate=pgo_prof.dir pgo.cpp -o pgo
```

The -fprofile-generate option enables instrumentation-based PGO. The path that we added after this flag is the directory where profiling data will be stored.

2. Next, we must run the pgo program with three command-line arguments:

```
$ ./pgo `seq 1 3`
Hello 0
Hello 6
...
Hello 36
Hello 39
$
```

You might get a totally different output since there is only a 50% of chance of the string being printed.

After this, the pgo_prof.dir folder should contain the default_<hash>_<n>.profraw file, as shown here:

```
$ ls pgo_prof.dir
default_10799426541722168222_0.profraw
```

The *hash* in the filename is a hash that's calculated based on your code.

3. We cannot directly use the `*.profraw` file for our second compilation phase. Instead, we must convert it into another kind of binary form using the `llvm-profdata` tool. Here is the command:

```
$ llvm-profdata merge pgo_prof.dir/ -o pgo_prof.profdata
```

`llvm-profdata` is a powerful tool for inspecting, converting, and merging profiling data files. We will look at it in more detail later. In the preceding command, we are merging and converting all the data files under `pgo_prof.dir` into a *single* `*.profdata` file.

4. Finally, we can use the file we just merged for the second stage of compilation. Here is the command:

```
$ clang -O1 -fprofile-use=pgo_prof.profdata pgo.cpp \
         -emit-llvm -S -o pgo.after.ll
```

Here, the `-fprofile-use` option told `clang` to use the profiling data stored in `pgo_prof.profdata` to optimize the code. We are going to look at the LLVM IR code after we've done this optimization.

Open `pgo.after.ll` and navigate to the `foo` function. Here is a simplified version of `foo`:

```
define void @foo(i32 %x) !prof !71 {
entry:
  %call = call i32 @get_random()
  %cmp = icmp sgt i32 %call, 5
  br i1 %cmp, label %if.then, label %if.end, !prof !72

if.then:
  %mul = mul nsw i32 %x, 3
  ...
}
```

In the preceding LLVM IR code, two places were different from the original IR; that is, the `!prof` tags that followed after both the function header and the branch instruction, which correspond to the `if(get_random() > 5)` code we saw earlier.

In LLVM IR, we can attach **metadata** to different IR units to provide supplementary information. Metadata will show up as a tag starting with exclamation mark (`'!'`) in the textual LLVM IR. `!prof`, `!71`, and `!72` in the preceding code are metadata tags that represent the profiling data we collected. More specifically, if we have profiling data associated with an IR unit, it always starts with `!prof`, followed by another metadata tag that contains the required values. These metadata values are put at the very bottom of the IR file. If we navigate there, we will see the content of `!71` and `!72`. Here is the code:

```
!71 = !{!"function_entry_count", i64 110}
!72 = !{!"branch_weights", i32 57, i32 54}
```

These two metadata are tuples with two and three elements. `!71`, as suggested by its first element, represents the number of times the `foo` function was called (in this case, it was called 110 times).

On the other hand, `!72` marks the number of times each branch in the `if (get_random() > 5)` statement was taken. In this case, the true branch was taken 57 times and the false branch was taken 54 times. We got these numbers because we were using uniform distribution for random number generation (namely, a ~50% chance for each branch).

In the second part of this section, we will learn how to access these values for the sake of developing a more aggressive compiler optimization. Before we do that, though, let's take a deeper look at the profiling data file we just collected.

The `llvm-profdata` tool we just used can not only help us convert the format of the profiling data, but also gives us a quick preview of its content. The following command prints out the summary for `pgo_prof.profdata`, including the profiling values that were collected from every function:

```
$ llvm-profdata show --all-functions --counts pgo_prof.profdata
...
  foo:
    Hash: 0x0ae15a44542b0f02
    Counters: 2
    Block counts: [54, 57]
  main:
    Hash: 0x0209aa3e1d398548
    Counters: 2
    Block counts: [110, 1]
...
```

```
Instrumentation level: IR  entry_first = 0
Functions shown: 9
Total functions: 9
Maximum function count: …
Maximum internal block count: …
```

Here, we can see the profiling data entries for each function. Each entry has a list of numbers representing the *execution frequency* of all the enclosing basic blocks.

Alternatively, you can inspect the same profiling data file by converting it into a textual file first. Here is the command:

```
$ llvm-profdata merge --text pgo_prof.profdata -o pgo_prof.
proftext
$ cat pgo_prof.proftext
# IR level Instrumentation Flag
:ir
…
foo
# Func Hash:
784007059655560962
# Num Counters:
2
# Counter Values:
54
57
…
```

The `*.proftext` file is in a human-readable textual format where all the profiling data is simply put in its own line.

This textual representation can actually be converted *back* into the `*.profdata` format using a similar command. Here is an example:

```
$ llvm-profdata merge --binary pgo_prof.proftext -o pgo_prof.
profdata
```

Therefore, `*.proftext` is especially useful when you want to *edit* the profiling data manually.

Before we dive into the APIs for PGO, there is one more concept we need to learn about: the instrumentation level.

Understanding the instrumentation level

So far, we've learned that instrumentation-based PGO can insert instrumentation code for collecting runtime profiling data. On top of this fact, the places where we insert this instrumentation code and its granularity also matter. This property is called the **instrumentation level** in instrumentation-based PGO. LLVM currently supports three different instrumentation levels. Here are descriptions of each:

* **IR**: Instrumentation code is inserted based on LLVM IR. For example, the code to collect the number of taken branches is directly inserted before a branch instruction. The `-fprofile-generate` command-line option we introduced earlier will generate profiling data with this instrumentation level. For example, let's say we have the following C code:

```
void foo(int x) {
  if (x > 10)
    puts("hello");
  else
    puts("world");
}
```

The corresponding IR – without enabling instrumentation-based PGO – is shown here:

```
define void @foo(i32 %0) {
  ...
  %4 = icmp sgt i32 %3, 10
  br i1 %4, label %5, label %7

5:
  %6 = call i32 @puts(..."hello"...)
  br label %9
7:
  %8 = call i32 @puts(..."world"...)
  br label %9

9:
  ret void
}
```

As we can see, there is a branch to either basic block; that is, %5 or %7. Now, let's generate the IR with instrumentation-based PGO enabled with the following command:

```
$ clang -fprofile-generate -emit-llvm -S input.c
```

This is the same command we used for the first PGO compilation phase in the *Working with profiling data* section, except that we are generating LLVM IR instead of an executable. The resulting IR is shown here:

```
define void @foo(i32 %0) {
    ...
    %4 = icmp sgt i32 %3, 10
    br i1 %4, label %5, label %9


5:
    %6 = load i64, i64* ... @__profc_foo.0, align 8
    %7 = add i64 %6, 1
    store i64 %7, i64* ... @__profc_foo.0, align 8
    %8 = call i32 @puts(..."hello"...)
    br label %13
9:
    %10 = load i64, i64* ... @__profc_foo.1, align 8
    %11 = add i64 %10, 1
    store i64 %11, i64* ... @__profc_foo.1, align 8
    %12 = call i32 @puts(..."world"...)
    br label %13


13:
    ret void
}
```

In the preceding IR, the basic blocks in both branches have new code in them. More specifically, both increment the value in a global variable – either @__profc_foo.0 or @__profc_foo.1 – by one. The values in these two variables will eventually be exported as the profiling data for branches, representing the number of times each branch was taken.

This instrumentation level provides decent precision but suffers from compiler changes. More specifically, if Clang changes the way it emits LLVM IR, the places where the instrumentation code will be inserted will also be different. This effectively means that for the same input code, the profiling data that's generated with an older version of LLVM might be incompatible with the profiling data that's generated with a newer LLVM.

- **AST**: Instrumentation code is inserted based on an AST. For example, the code to collect the number of taken branches might be inserted as a new `Stmt` AST node inside an `IfStmt` (an AST node). With this method, the instrumentation code is barely affected by compiler changes and we can have a more stable profiling data format across different compiler versions. The downside of this instrumentation level is that it has less precision than the IR instrumentation level.

  You can adopt this instrumentation level by simply using the `-fprofile-instr-generate` command-line option in place of `-fprofile-generate` when invoking `clang` for the first compilation. You don't need to change the command for the second compilation, though.

- **Context-sensitive**: In the *Working with profiling data* section, we learned that we could collect information about the number of times a branch has been taken. However, with the IR instrumentation level, it's nearly impossible to tell which *caller function* leads to the most branch that has been taken the most. In other words, the conventional IR instrumentation level loses the calling *context*. The context-sensitive instrumentation level tries to address this problem by collecting the profiles again once the functions have been inlined, thus creating profiling data with higher precision.

  However, it's slightly cumbersome to use this feature in Clang – we need to compile the same code *three* times rather than twice. Here are the steps for using context-sensitive PGO:

```
$ clang -fprofile-generate=first_prof foo.c -o foo_exe
$ ./foo_exe …
$ llvm-profdata merge first_prof -o first_prof.profdata
```

First, generate some normal (IR instrumentation level) profiling data using what we learned in the *Working with profiling data* section:

```
$ clang -fprofile-use=first_prof.profdata \
        -fcs-profile-generate=second_prof foo.c -o foo_exe2
$ ./foo_exe2
$ llvm-profdata merge first_prof.profdata second_prof \
                    -o combined_prof.profdata
```

Then, run `clang` with two PGO command-line options, `-fprofile-use` and `-fcs-profile-generate`, with the path to the profiling file from the previous step and the prospective output path, respectively. When we use `llvm-profdata` to do the post-processing, we are merging all the profiling data files we have:

```
$ clang -fprofile-use=combined_prof.profdata \
        foo.c -o optimized_foo
```

Finally, feed the combined profiling file into Clang so that it can use this context-sensitive profiling data to get a more accurate portrait of the program's runtime behavior.

Note that different instrumentation levels only affect the accuracy of the profiling data; they don't affect how we *retrieve* this data, which we are going to talk about in the next section.

In the last part of this section, we are going to learn how to access this profiling data inside an LLVM pass via the APIs provided by LLVM.

Learning about the APIs for accessing profiling data

In the previous section, we learned how to run the instrumentation-based PGO using Clang and view the profiling data file using `llvm-profdata`. In this section, we are going to learn how to access that data within an LLVM pass to help us develop our own PGO.

Before we go into the development details, let's learn how to *consume* those profiling data files into `opt`, since it's easier to test individual LLVM pass using it. Here is a sample command:

```
$ opt -pgo-test-profile-file=pgo_prof.profdata \
      --passes="pgo-instr-use,my-pass…" pgo.ll …
```

There are two keys in the preceding command:

- Use `-pgo-test-profile-file` to designate the profiling data file you want to put in.

- The "pgo-instr-use" string represents the `PGOInstrumentaitonUse` pass, which reads (instrumentation-based) profiling files and *annotates* the data on an LLVM IR. However, it is not run by default, even in the predefined optimization levels (that is, O0 ~ O3, Os, and Oz). Without this pass being ahead in the Pass pipeline, we are unable to access any profiling data. Therefore, we need to explicitly add it to the optimization pipeline. The preceding sample command demonstrated how to run it before a custom LLVM pass, `my-pass`, in the pipeline. If you want to run it before any of the predefined optimization pipelines – for instance, O1 – you must specify the `--passes="pgo-instr-use,default<O1>"` command-line option.

Now, you might be wondering, what happens after the profiling data is read into `opt`? It turns out that the LLVM IR file that was generated by the *second* compilation phase – `pgo.after.ll` – has provided us with some answers to this question.

In `pgo.after.ll`, we saw that some branches were *decorated* with **metadata** specifying the number of times they were taken. Similar metadata appeared in functions, which represented the total number of times those functions were called.

More generally speaking, LLVM directly *combines* the profiling data – read from the file – with its associated IR constructions via **metadata**. The biggest advantage of this strategy is that we don't need to carry the raw profiling data throughout the entire optimization pipeline – the IR itself contains this profiling information.

Now, the question becomes, how can we access metadata that's been attached to IR? LLVM's metadata can be attached to many kinds of IR units. Let's take a look at the most common one first: accessing metadata attached to an `Instruction`. The following code shows us how to read the profiling metadata – `!prof !71`, which we saw previously – that's attached to a branch instruction:

```
// `BB` has the type of `BasicBlock&`
Instruction *BranchInst = BB.getTerminator();
MDNode *BrWeightMD = BranchInst->getMetadata(LLVMContext::MD_prof);
```

In the preceding snippet, we are using `BasicBlock::getTerminator` to get the last instruction in a basic block, which is a branch instruction most of the time. Then, we tried to retrieve the profiling metadata with the `MD_prof` metadata. `BrWeightMD` is the result we are looking for.

The type of `BrWeightMD`, MDNode, represents a single metadata node. Different MDNode instances can be *composed* together. More specifically, a MDNode instance can use other MDNode instances as its operands – similar to the `Value` and `User` instances we saw in *Chapter 10, Processing LLVM IR*. The compound MDNode can express more complex concepts.

For example, in this case, each operand in `BrWeightMD` represents the number of times each branch was taken. Here is the code to access them:

```
if (BrWeightMD->getNumOperands() > 2) {
  // Taken counts for true branch
  MDNode *TrueBranchMD = BrWeightMD->getOperand(1);
  // Taken counts for false branch
  MDNode *FalseBranchMD = BrWeightMD->getOperand(2);
}
```

As you can see, the taken counts are also expressed as MDNode.

> **Operand indices for both branches**
>
> Note that the data for both branches is placed at the operands starting from index 1 rather than index 0.

If we want to convert these branch MDNode instances into constants, we can leverage a small utility provided by the `mdconst` namespace. Here is an example:

```
if (BrWeightMD->getNumOperands() > 2) {
  // Taken counts for true branch
  MDNode *TrueBranchMD = BrWeightMD->getOperand(1);
  ConstantInt *NumTrueBrTaken
    = mdconst::dyn_extract<ConstantInt>(TrueBranchMD);
  ...
}
```

The previous code *unwrapped* an MDNode instance and extracted the underlying `ConstantInt` instance.

For `Function`, we can get the number of times it was called in an even easier way. Here is the code:

```
// `F` has the type of `Function&`
Function::ProfileCount EntryCount = F.getEntryCount();
uint64_t EntryCountVal = EntryCount.getCount();
```

`Function` is using a slightly different way to present its called frequency. But retrieving the numerical profiling value is still pretty easy, as shown in the preceding snippet.

It is worth noting that although we only focused on instrumentation-based PGO here, for *sampling*-based PGO, LLVM also uses the same programming interface to expose its data. In other words, even if you're using profiling data that's been collected from sampling tools with a different `opt` command, the profiling data will also be annotated on IR units and you can still access it using the aforementioned method. In fact, the tools and APIs we are going to introduce in the rest of this section are mostly profiling-data-source agnostic.

So far, we have been dealing with real values that have been retrieved from the profiling data. However, these low-level values cannot help us go far in terms of developing a compiler optimization or program analysis algorithm – usually, we are more interested in high-level concepts such as "functions that are executed *most frequently*" or "branches that are *least taken*". To address these demands, LLVM builds several analyses on top of profiling data to deliver such high-level, structural information.

In the next section, we are going to introduce some of these analyses and their usages in LLVM Pass.

Using profiling data analyses

In this section, we are going to learn three analysis classes that can help us reason about the execution frequency of basic blocks and functions at runtime. They are as follows:

- `BranchProbabilityInfo`
- `BlockFrequencyInfo`
- `ProfileSummaryInfo`

This list is ordered by their analyzing scope in the IR – from local to global. Let's start with the first two.

Using BranchProbabilityInfo and BlockFrequencyInfo

In the previous section, we learned how to access the profiling metadata that's attached to each branch instruction – the **branch weight**. The analysis framework in LLVM provides you with an even easier way to access these values via the `BranchProbabilityInfo` class. Here is some example code showing how to use it in a (function) Pass:

```
#include "llvm/Analysis/BranchProbabilityInfo.h"
PreservedAnalyses run(Function &F, FunctionAnalysisManager
&FAM) {
  BranchProbabilityInfo &BPI
    = FAM.getResult<BranchProbabilityAnalysis>(F);
  BasicBlock *Entry = F.getEntryBlock();
  BranchProbability BP = BPI.getEdgeProbability(Entry, 0);
  …
}
```

The previous code retrieved a `BranchProbabilityInfo` instance, which is the result of `BranchProbabilityAnalysis`, and tried to get the weight from the entry block to its first successor block.

The returned value, a `BranchProbability` instance, gives you the branch's probability in the form of a percentage. You can use `BranchProbability::getNumerator` to retrieve the value (the "denominator" is 100 by default). The `BranchProbability` class also provides some handy utility methods for performing arithmetic between two branch probabilities or scaling the probability by a specific factor. Although we can easily tell which branch is more likely to be taken using `BranchProbabilityInfo`, without additional data, we can't tell the branch's probability (to be taken) in the *whole function*. For example, let's assume we have the following CFG:

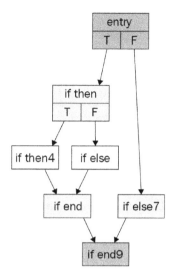

Figure 12.6 – CFG with nested branches

For the preceding diagram, we have profiling counter values for the following basic blocks:

- **if.then4**: 2
- **if.else**: 10
- **if.else7**: 20

If we *only* look at the branch weight metadata toward blocks if.then4 and if.else – that is, the true and false branches for if.then, respectively – we might come under the *illusion* that the if.else block has a ~83% chance of being taken. But the truth is, it only has a ~31% chance because the control flow has a higher probability to go into if.else7 before even entering the if.then region. Of course, in this case, we can do simple math to figure out the correct answer, but when the CFG is getting bigger and more complex, we might have a hard time doing this by ourselves.

The BlockFrequencyInfo class provides a shortcut to this problem. It can tell us the frequency of each basic block to be taken under the context of its enclosing function. Here is an example of its usage in a Pass:

```
#include "llvm/Analysis/BlockFrequencyInfo.h"
PreservedAnalyses run(Function &F, FunctionAnalysisManager
&FAM) {
  BlockFrequencyInfo &BFI
    = FAM.getResult<BlockFrequencyAnalysis>(F);
```

```
   for (BasicBlock *BB : F) {
     BlockFrequency BF = BFI.getBlockFreq(BB);
   }
   ...
 }
```

The previous code retrieved a `BlockFrequencyInfo` instance, which is the result of `BlockFrequencyAnalysis`, and tried to evaluate the block frequency of each basic block in the function.

Similar to the `BranchProbability` class, `BlockFrequency` also provides nice utility methods to calculate with other `BlockFrequency` instances. But different from `BranchProbability`, the numeric value that's retrieved from `BlockFrequency` is not presented as a percentage. More specifically, `BlockFrequency::getFrequency` returns an integer that is the frequency *relative* to the entry block of the current function. In other words, to get a percentage-based frequency, we can use the following snippet:

```
// `BB` has the type of `BasicBlock*`
// `Entry` has the type of `BasicBlock*` and represents entry
// block
BlockFrequency BBFreq = BFI.getBlockFreq(BB),
               EntryFreq = BFI.getBlockFreq(Entry);
auto FreqInPercent
   = (BBFreq.getFrequency() / EntryFreq.getFrequency()) * 100;
```

The highlighted `FreqInPercent` is the block frequency of BB, expressed as a percentage.

`BlockFrequencyInfo` calculates the frequency of a specific basic block under the context of a function – but what about the entire *module*? More specifically, if we bring a **call graph** into this situation, can we multiply the block frequency we just learned with the frequency of the enclosing function being called, in order to get the execution frequency in the *global* scope? Fortunately, LLVM has prepared a useful class to answer this question – `ProfileSummaryInfo`.

Using ProfileSummaryInfo

The `ProfileSummaryInfo` class gives you a global view of all the profiling data in a `Module`. Here is an example of retrieving an instance of it inside a module Pass:

```
#include "llvm/Analysis/ProfileSummaryInfo.h"
PreservedAnalyses run(Module &M, ModuleAnalysisManager &MAM) {
```

```
ProfileSummaryInfo &PSI = MAM.
getResult<ProfileSummaryAnalysis>(M);

...

}
```

`ProfileSummaryInfo` provides a wide variety of functionalities. Let's take a look at three of its most interesting methods:

- `isFunctionEntryCold/Hot(Function*)`: These two methods compare the entry count of a `Function` – which effectively reflects the number of times a function was called –against that of other functions in the same module and tell us if the inquiry function is ranking high or low in this metric.

- `isHot/ColdBlock(BasicBlock*, BlockFrequencyInfo&)`: These two methods work similarly to the previous bullet one but compare the execution frequency of a `BasicBlock` against *all* the other blocks in the module.

- `isFunctionCold/HotInCallGraph(Function*, BlockFrequencyInfo&)`: These two methods combine the methods from the previous two bullet points they can tell you whether a function is considered hot or cold based on its entry count or the execution frequency of its enclosing basic blocks. This is useful when a function has a low entry count – that is, it was not called often – but contains a *loop* that has extremely a high basic block execution frequency. In this case, the `isFunctionHotInCallGraph` method can give us a more accurate assessment.

These APIs also have variants where you can designate the cutoff point as being "hot" or "cold." Please refer to the API documentation for more information.

For a long time, the compiler was only able to analyze and optimize the source code with a static view. For dynamic factors inside a program – for instance, the branch taken count – compilers could only make an approximation. PGO opened an alternative path to provide extra information for compilers to peek into the target program's runtime behavior, for the sake of making less ambiguous and more aggressive decisions. In this section, we learned how to collect and use runtime profiling information – the key to PGO – with LLVM. We learned how to use the related infrastructure in LLVM to collect and generate such profiling data. We also learned about the programming interface we can use to access that data – as well as some high-level analyses built on top of it – to assist our development inside an LLVM Pass. With these abilities, LLVM developers can plug in this runtime information to further improve the quality and precision of their existing optimization Passes.

Summary

In this chapter, we *augmented* the workspace of the compiler by processing the static source code and capturing the program's runtime behaviors. In the first part of this chapter, we learned how to use the infrastructure provided by LLVM to create a sanitizer – a technique that inserts instrumentation code into the target program for the sake of checking certain runtime properties. By using a sanitizer, software engineers can improve their development quality with ease and with high precision. In the second part of this chapter, we extended the usages of such runtime data to the domain of compiler optimization; PGO is a technique that uses dynamic information, such as the execution frequency of basic blocks or functions, to make more aggressive decisions for optimizing the code. Finally, we learned how to access such data with an LLVM Pass, which enables us to add PGO enhancement to existing optimizations.

Congratulations, you've just finished the last chapter! Thank you so much for reading this book. Compiler development has never been an easy subject – if not an obscure one – in computer science. In the past decade, LLVM has significantly lowered the difficulties of this subject by providing robust yet flexible modules that fundamentally change how people think about compilers. A compiler is not just a single executable such as `gcc` or `clang` anymore – it is a collection of building blocks that provide developers with *countless* ways to create tools to deal with hard problems in the programming language field.

However, with so many choices, I often became lost and confused when I was still a newbie to LLVM. There was documentation for every single API in this project, but I had no idea how to put them together. I wished there was a book that pointed in the general direction of each important component in LLVM, telling me *what* it is and *how* I can take advantage of it. And here it is, the book I wished I could have had at the beginning of my LLVM career – the book you just finished – come to life. I hope you won't stop your expedition of LLVM after finishing this book. To improve your skills even further and reinforce what you've learned from this book, I recommend you to check out the official document pages (`https://llvm.org/docs`) for content that complements this book. More importantly, I encourage you to participate in the LLVM community via either their mailing list (`https://lists.llvm.org/cgi-bin/mailman/listinfo/llvm-dev`) or Discourse forum (`https://llvm.discourse.group/`), especially the first one – although a mailing list might sound old-school, there are many talented people there willing to answer your questions and provide useful learning resources. Last but not least, annual LLVM dev meetings (`https://llvm.org/devmtg/`), in both the United States and Europe, are some of the best events where you can learn new LLVM skills and chat face-to-face with people who literally built LLVM.

I hope this book enlightened you on your path to mastering LLVM and helped you find joy in crafting compilers.

Assessments

This section contains answers to the questions from all chapters.

Chapter 6, Extending the Preprocessor

1. Most of the time tokens are harvested from the provided source code, but in some cases, tokens might be generated dynamically inside the Preprocessor. For example, the __LINE__ built-in macro is expanded to the current line number, and the __DATE__ macro is expanded to the current calendar date. How does Clang put that generated textual content into the SourceManager's source code buffer? How does Clang assign SourceLocation to these tokens?

-   Developers can leverage the clang::ScratchBuffer class to insert dynamic Token instances.

2. When we were talking about implementing a custom PragmaHandler, we were using Preprocessor::Lex to fetch tokens followed after the pragma name, until we hit the eod token type. Can we keep lexing *beyond* the eod token? What interesting thing will you do if you can consume arbitrary tokens follow after the #pragma directive?

-   Yes, we can keep lexing beyond the eod token. It simply consumes the contents following the #pragma line. In this way, you can create a custom #pragma that allows you to write *arbitrary* content (below it) – for instance, writing programming languages that are not supported by Clang. Here is an example:

```
#pragma javascript function
const my_func = (arg) => {
  console.log(`Hello ${arg}`);
};
```

The preceding snippet showed how to create a custom #pragma that allows you to define a JavaScript function below it.

3. In the `macro guard` project from the *Developing custom preprocessor plugins and callbacks* section, the warning message has the format of [WARNING] In <source location>: ..... Apparently, this is not the typical compiler warning we see from Clang, which looks like <source location>: warning: ...:

    ```
    ./simple_warn.c:2:7: warning: unused variable 'y'...
      int y = x + 1;
          ^

    1 warning generated.
    ```

    The `warning` string is even colored in supported terminals. How can we print a warning message like that? Is there an infrastructure in Clang for doing that?

    *   Developers can use the diagnostics framework in Clang to print messages like this. In the *Printing diagnostics messages* section of *Chapter 7, Handling AST*, we will show you some of the usages of this framework.

Chapter 8, Working with Compiler Flags and Toolchains

1. It is common to override the assembling and linking stage, since different platforms tend to support different assemblers and linkers. But is it possible to override the *compiling* stage (which is Clang)? If it is possible, how do we do it? What might be the possible reasons for people to do that?

    *   You can override the `ToolChain::SelectTool` method and provide an alternative `Tool` instance (which represents the compilation stage) according to the argument. Here is an example:

    ```
    Tool*
    MyToolChain::SelectTool(const JobAction &JA) const
    override {
      if (JA.getKind() == Action::CompileJobClass &&
          getTriple().getArch() == CUSTOM_HARDWARE)
        return new MyCompiler(...);

      ...
      // Run the default `SelectTool` otherwise
      return ToolChain::SelectTool(JA);
    }
    ```

In the preceding snippet, we provided our own compiler instance – `MyCompiler` – which is a class derived from `Tool`, if we are trying to compile the code for a certain hardware architecture.

Providing an alternative compiler instance is useful when your target platform (for example, the `CUSTOM_HARDWARE` in the preceding snippet) or input file is not supported by Clang, but you still want to use the *same* `clang` command-line interface for all the build jobs. For example, suppose you are trying to cross-compile the same projects to *multiple* different architectures, but some of them are not supported by Clang yet. Therefore, you can create a custom Clang toolchain and redirect the compilation job to an external compiler (for example, `gcc`) when the `clang` command-line tool is asked to build the project for those architectures.

2. When we were working on `tools::zipline::Linker::ConstructJob`, we simply use `llvm_unreachable` to bail out the compilation process if the user provides an unsupported compressor name through the `-fuse-ld` flag. Can we replace it with Clang's **diagnostic** framework that we learned about in *Chapter 7, Handling AST*, to print out better messages?

- The `Driver` class provides a shortcut to access the diagnostic framework. Inside a derived class of `Tool`, you can use `getToolChain().getDriver()` to get a `Driver` instance, then print out the diagnostic message using the `Driver::Diag` method.

3. Just like we can use `-Xclang` to pass flags directly to the frontend, we can also pass assembler-specific or linker-specific flags directly to the assembler or linker using driver flags such as `-Wa` (for assembler) and `-Wl` (for linker). How do we consume those flags in our custom assembler and linker stages within Zipline?

- Inside the `ConstructJob` method, you can read the value of `options::OPT_Wa_COMMA` and `options::OPT_Wl_COMMA` to retrieve assembler- and linker-specific command line flags, respectively. Here is an example:

```
void
MyAssembler::ConstructJob(Compilation &C,
                          const JobAction &JA,
                          const InputInfo &Output,
                          const InputInfoList &Inputs,
                          const ArgList &Args,
                          const char *LinkingOutput)
                          const {
```

```
if (Arg *A = Args.getLastArg(options::OPT_Wl_COMMA)) {
  // `A` contains linker-specific flags
  …
}
…
}
```

Chapter 9, Working with PassManager and AnalysisManager

1. In the StrictOpt example in the *Writing a LLVM Pass for the new PassManager* section, how do we write a Pass without deriving the `PassInfoMixin` class?

*   The `PassInfoMixin` class only defines a utility function for you, `name`, which returns the name of this Pass. Therefore, you can easily create one by yourself. Here is an example:

```
struct MyPass {
  static StringRef name() { return "MyPass"; }
  PreservedAnalyses run(Function&,
FunctionAnalysisManager&);
};
```

2. How do we develop custom instrumentation for the new PassManager? How do we do it without modifying the LLVM source tree? (Hint: Use the Pass plugin we learned in this chapter.)

*   Pass instrumentation is a piece of code that runs before and/or after an LLVM Pass. This blog post shows an example of developing a custom Pass instrumentation via the Pass plugin: https://medium.com/@mshockwave/writing-pass-instrument-for-llvm-newpm-f17c57d3369f.

Other Books You May Enjoy

If you enjoyed this book, you may be interested in these other books by Packt:

Getting Started with LLVM Core Libraries

Bruno Cardoso Lopes, Rafael Auler

ISBN: 9781782166924

- Configure, build, and install extra LLVM open source projects including Clang tools, static analyzer, Compiler-RT, LLDB, DragonEgg, libc++, and LLVM test-suite

- Understand the LLVM library design and interaction between libraries and standalone tools

- Increase your knowledge of source code processing stages by learning how the Clang frontend uses a lexer, parser, and syntax analysis

- Manipulate, generate, and play with LLVM IR files while writing custom IR analyses and transformation passes

- Write tools to use LLVM Just-in-Time (JIT) compilation capabilities

- Find bugs and improve your code by using the static analyzer

- Design source code analysis and transformation tools using LibClang, LibTooling, and the Clang plugin interface

Learn LLVM 12

Kai Nacke

ISBN: 9781839213502

- Configure, compile, and install the LLVM framework
- Understand how the LLVM source is organized
- Discover what you need to do to use LLVM in your own projects
- Explore how a compiler is structured, and implement a tiny compiler
- Generate LLVM IR for common source language constructs
- Set up an optimization pipeline and tailor it for your own needs
- Extend LLVM with transformation passes and clang tooling
- Add new machine instructions and a complete backend

Packt is searching for authors like you

If you're interested in becoming an author for Packt, please visit authors. packtpub.com and apply today. We have worked with thousands of developers and tech professionals, just like you, to help them share their insight with the global tech community. You can make a general application, apply for a specific hot topic that we are recruiting an author for, or submit your own idea.

Leave a review - let other readers know what you think

Please share your thoughts on this book with others by leaving a review on the site that you bought it from. If you purchased the book from Amazon, please leave us an honest review on this book's Amazon page. This is vital so that other potential readers can see and use your unbiased opinion to make purchasing decisions, we can understand what our customers think about our products, and our authors can see your feedback on the title that they have worked with Packt to create. It will only take a few minutes of your time, but is valuable to other potential customers, our authors, and Packt. Thank you!

Index

U

Uniform Resource Locator (URL) 277

V

values
  working with 229-234

W

worklist 242
workloads 149

Y

YAML Ain't Markup Language
    (YAML) 265

Z

Zipline 164

www.ingramcontent.com/pod-product-compliance
Lightning Source LLC
Chambersburg PA
CBHW062050050326
40690CB00016B/3047